Weak Interactions in Nuclei

Princeton Series in Physics

edited by Arthur S. Wightman
and Philip W. Anderson

Quantum Mechanics for Hamiltonians Defined as Quadratic Forms *by Barry Simon*

Lectures on Current Algebra and Its Applications *by Sam B. Treiman, Roman Jackiw, and David J. Gross*

Physical Cosmology *by P.J.E. Peebles*

The Many-Worlds Interpretation of Quantum Mechanics
edited by B. S. DeWitt and N. Graham

The $P(\Phi)_2$ Euclidean (Quantum) Field Theory *by Barry Simon*

Homogeneous Relativistic Cosmologies *by Michael P. Ryan, Jr. and Lawrence C. Shepley*

Studies in Mathematical Physics: Essays in Honor of Valentine Bargmann
edited by Elliott H. Lieb. B. Simon, and A. S. Wightman

Convexity in the Theory of Lattice Gases *by Robert B. Israel*

Surprises in Theoretical Physics *by Rudolf Peierls*

Works on the Foundations of Statistical Physics *by N. S. Krylov*

The Large-Scale Structure of the Universe *by P.J.E. Peebles*

Quantum Theory and Measurement
edited by John Archibald Wheeler and Wojciech Hubert Zurek

Statistical Physics and the Atomic Theory of Matter, from Boyle and Newton to Landau and Onsager *by Stephen G. Brush*

Supersymmetry and Supergravity *by J. Wess and J. Bagger*

Current Algebra and Anomalies *by Sam B. Treiman, Roman Jackiw, Bruno Zumino, and Edward Witten*

Spin Glasses and Other Frustrated Systems *by D. Chowdhury*

Weak Interactions in Nuclei *by Barry R. Holstein*

Weak Interactions in Nuclei

Barry R. Holstein

Princeton Series in Physics

Princeton University Press
Princeton, New Jersey

Published by Princeton University Press,
41 William Street, Princeton, New Jersey 08540
In the United Kingdom:
Princeton University Press, Guildford, Surrey

Library of Congress Cataloging-in-Publication Data

Holstein, Barry, 1943–
Weak interactions in nuclei.

(Princeton series in physics)
Includes index.
1. Nuclear physics. 2. Weak interactions (Nuclear physics)
I. Title. II. Series.
QC776.H58 1989 539.7′54 88-25451
ISBN 0-691-08523-4

This book has been composed in Times Roman

Clothbound editions of Princeton University Press books are printed on acid-free paper, and binding materials are chosen for strength and durability.

Printed in the United States of America
by Princeton University Press,
Princeton, New Jersey

To C, J, and J

CONTENTS

PREFACE

In the summer of 1985 I delivered a set of ten hour-long lectures to a group of graduate students and postdocs in nuclear physics as part of the first Georgetown University summer school. The following autumn I was a visitor at Princeton University and presented an expanded version of these talks to a group of graduate students and nuclear physics faculty. My purpose was not to give a complete discussion of any particular aspect of research but rather to present a wide overview, emphasizing areas where weak interaction studies using the nucleus as a laboratory are contributing at the cutting edge of knowledge.

I attempted to pitch my presentation at the advanced graduate student/postdoc level, assuming a familiarity with quantum mechanics and basic nuclear physics but not of detailed concepts in particle physics. For that reason, I have given in chapter 2 a simple semihandwaving approach to the physics of the quark model and in chapter 3 a brief derivation of the Weinberg-Salam model. Those familiar with this theoretical background may well wish to skip these sections. The primary focus of my discussion is presented in the remaining chapters, where I discuss experimental work involving the nuclear medium which is relevant to this picture.

When Princeton University Press suggested that I prepare these notes into a monograph, I considered making my discussion more authoritative and complete. However, I soon rejected that idea. Rather, I have attempted to retain much of the original informality which characterizes the presentation at the summer school. I hope thereby to offer the message that my discussion is not at all the final word but rather represents only an introduction, which invites the reader to become a participant in this search.

AMHERST, MASS.
AUGUST 1988

ACKNOWLEDGMENTS

It is a pleasure to acknowledge the contributions of those who helped to make this book possible: Paul Trudeau, whose summer program led to the development of these lectures; Frank Calaprice, whose invitation allowed me to turn these ideas into prose; the National Science Foundation for their financial support; and, finally, Mrs. Nellie Bristol, who returned even after her retirement to complete the typing of the manuscript.

Weak Interactions
in Nuclei

Chapter 1

INTRODUCTION

I wish to begin this monograph with the clear statement that my training is as a particle physicist. And yet the lectures on which this text is based were presented to audiences consisting almost entirely of nuclear physicists. In previous eras, this scenario might have been unlikely, as particle and nuclear physics were considered as quite disparate disciplines, with little overlap between them. On the other hand, during the present decade a healthy symbiotic relationship has developed, with ideas from particle physics—quark structure, solitons, etc.—playing an increasing role in nuclear physics and experiments performed using nuclei—on axions, T-violation, etc.—probing important aspects of the fundamental interactions. It is precisely this sort of "intersection" physics that I intend to discuss, using the weak interaction as a focus. Much of my discussion will involve aspects of basic symmetries that are tested using the nucleus as a laboratory. Such tests are usefully performed in nuclei since, involving universal symmetry propositions, they are to a great extent insulated from wave function uncertainties that often plague other nuclear experiments.

The concept of symmetry has always been fascinating to humankind. The Pythagorans considered the circle to be the most perfect of two-dimensional objects and the sphere to embody perfection in three dimensions because of their obvious symmetries. In this ancient world view the stars were fixed in the heavenly spheres and the planets were considered to move in perfect circles about the earth. Of course, it was soon realized that the outer planets—Mars, Jupiter, etc.—occasionally doubled back upon themselves during their wanderings through the heavens. This apparent breakdown of symmetry was repaired with the idea of epicycles—circles moving on circles (Figure 1.1). Even Kepler, whose careful analysis of Tycho's astronomical observations produced the first correct treatment of planetary motion, had his own picture of planetary structure based on symmetry, with interlocking circles and regular solids, as in a child's puzzle (Figure 1.2). In fact it was only with the realization by Isaac Newton that it is the physical *laws* not the *orbits* that are symmetric which enabled real progress in physics to take place.

The importance of symmetry considerations in field theory stems from a mathematical proposition called "Noether's theorem," which states that for every invariance of the Lagrangian or Hamiltonian of a system there

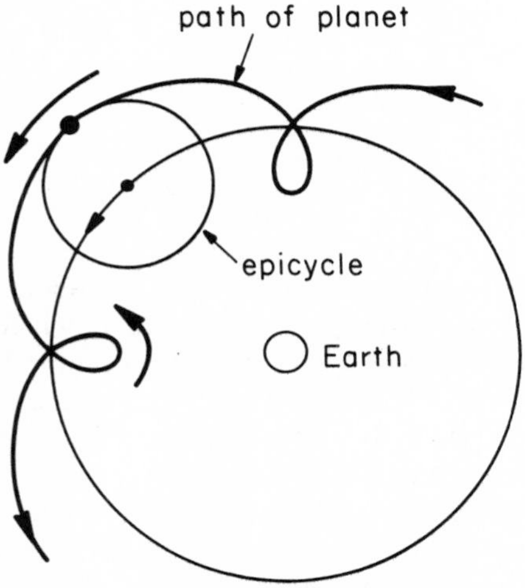

FIGURE 1.1. In order to explain the observation that the outer planets double back on themselves, the idea of epicycles—or circles moving on circles—was developed. Of course, this is only required in a geocentric view of planetary motion.

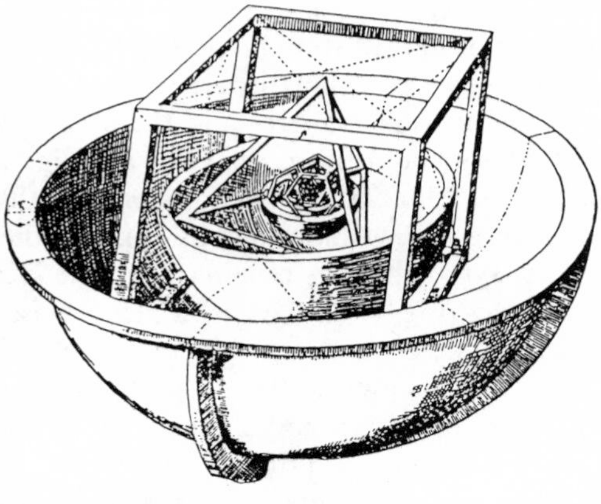

FIGURE 1.2. In Kepler's (1595) view, the orbits of the planets were associated with spheres that are separated by the regular solids—cube, tetrahedron, dodecahedron, octahedron and icosahedron. Kepler did not know of the outer three planets; indeed, Uranus was not discovered until 1781.

exists a corresponding conserved quantity. One has, for example,

Invariance	*Conservation Law*
Spatial translation	Momentum
Time translation	Energy
Rotation	Angular momentum
Reflection	Parity

etc.

Thus the existence of a symmetry can reveal a great deal concerning the structure of the underlying dynamics of a physical system, and that will prove to be the case for the symmetries analyzed below.

It is my hope that this discussion of symmetry tests involving the weak interaction within the nuclear medium will convey a strong sense of how nuclear physics has and is continuing to play an important role on the fundamental physics "frontier." Since this is a brief monograph—not a definitive treatise—and there is a great deal of material to be covered, I shall of necessity paint with the proverbial "broad brush." Also, since this is not intended as a detailed review article but is rather in the spirit of an introductory overview, I have not attempted to provide a complete bibliography but rather have chosen to cite representative works. (I apologize in advance to those authors whose papers are slighted and hope that they will sympathize with my purpose.) An outline of the material to be presented is given in the Table of Contents. A glance will reveal that the list of subjects is not intended to depict a complete picture of the state of the field but is rather a guided tour over a number of selected and, hopefully, representative topics.

Chapter 2

A NUCLEAR PHYSICIST'S GUIDE TO THE QUARK MODEL

"Three quarks for Muster Mark!"
JAMES JOYCE, *FINNEGAN'S WAKE*

Having dispensed with the mandatory introductory remarks, we can begin to examine some of the basic ideas of the quark model that all nuclear physicists in the 1980s should have at their disposal.

We start with atomic physics and review the idea that fundamental interactions are due to virtual particle exchange. Thus, for example, one can consider the electromagnetic interaction, which binds the hydrogen atom, from the point of view of either (1) potential scattering, or (2) virtual photon exchange. From the former perspective, the dominant piece of the e–p interaction has three key components [1]:

$$V(r) = V_0(r) + V_1(r) + V_2(r), \tag{2.1}$$

where

$$V_0(r) = -\frac{\alpha}{r} \tag{2.2}$$

is the usual Coulomb potential;

$$V_1(r) = -\frac{p^4}{8m_e^{\,3}} - \frac{e}{8m_e^{\,2}}\,\mathbf{\nabla}\cdot\mathbf{E} + \frac{1}{4m_e^{\,2}}\frac{1}{r}\frac{d}{dr}V_0\boldsymbol{\sigma}\cdot\mathbf{L} \tag{2.3}$$

is the piece of the interaction responsible for fine structure [2]; and

$$V_2(r) = (\boldsymbol{\mu}_e \times \mathbf{\nabla})\cdot(\boldsymbol{\mu}_p \times \mathbf{\nabla})\frac{1}{4\pi r} \tag{2.4}$$

is the spin-spin force that produces hyperfine [3] splitting. Here

$$\alpha = \frac{e^2}{4\pi} \cong \frac{1}{137} \tag{2.5}$$

is the fine structure constant, and

$$\boldsymbol{\mu}_e = \frac{e}{2m_e}\boldsymbol{\sigma}_e, \qquad \boldsymbol{\mu}_p = \frac{2.79e}{2m_p}\boldsymbol{\sigma}_p \tag{2.6}$$

are the magnetic moments of electron and proton, respectively.

Of course, in the hydrogen atom, these latter contributions represent quite small, but important, corrections to the basic energy levels:

$$V_0(r)\text{: } E_n^{(0)} = -\frac{\alpha^2}{2n^2} m_e \qquad n = 1, 2, 3, \ldots \tag{2.7}$$

determined by $V_0(r)$—

$$\begin{aligned} &V_1(r)\text{: } \frac{\Delta E_1(\text{fine structure})}{E^{(0)}} \sim \alpha^2 \sim 10^{-4} \\ &V_2(r)\text{: } \frac{\Delta E_2(\text{hyperfine structure})}{E^{(0)}} \sim \alpha^2 \frac{m_e}{m_p} \sim 10^{-7}. \end{aligned} \tag{2.8}$$

The standard textbook derivation of this effective e–p–potential is somewhat involved [1], discussing the physics motivation of each term separately. However, great simplicity and power result from the idea of the $e - p$ interaction arising from virtual photon exchange (Figure 2.1), as shown in the appendix. In this approach all terms arise naturally, with no need to append any pieces on an ad hoc basis.

Another interesting feature of atomic physics is that, while the fundamental Coulomb interaction between charges drops off as $1/r$, the force between two bound states—hydrogen atoms—has a completely different character. This residual "van der Waals" interaction behaves as $1/r^6$ and may be thought of in terms of the exchange of a pair of virtual photons

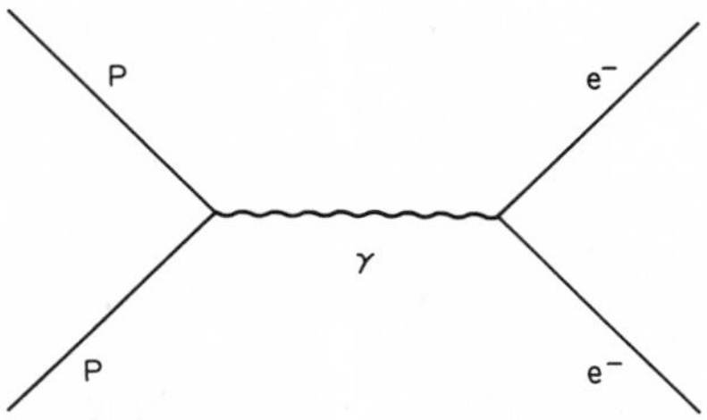

FIGURE 2.1. The Coulomb interaction between electron and proton takes place through the exchange of a virtual photon.

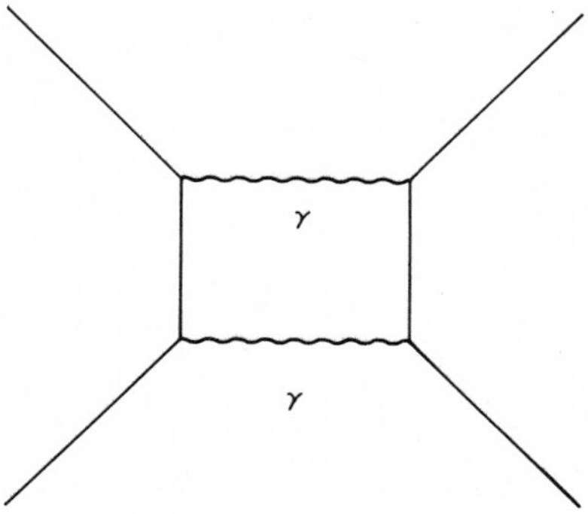

FIGURE 2.2. The Van der Waals interaction between a pair of neutral atoms can be considered in terms of the exchange of a pair of virtual photons.

between atoms, one to polarize the atom, the second to interact with the induced polarization (Figure 2.2) [4]. More quantitatively, the van der Waals interaction is a consequence of the familiar dipole-dipole potential

$$V_{12} = (\mathbf{p}_1 \times \nabla) \cdot (\mathbf{p}_2 \times \nabla) \frac{1}{4\pi r_{12}} \tag{2.9}$$

where

$$\mathbf{p} = \sum_i e_i \mathbf{r}_i \tag{2.10}$$

is the electric dipole moment operator. Symbolically we may write

$$V_{12} \sim \frac{\sum_{ij} e_i^{(1)} r_i^{(1)} e_j^{(2)} r_j^{(2)}}{r_{12}^3} \tag{2.11}$$

In first-order perturbation theory this effect vanishes

$$\langle V_{12} \rangle = 0 \tag{2.12}$$

since

$$\langle r_i \rangle = 0. \tag{2.13}$$

However, to second order we find a nonvanishing result,

$$V_{\text{eff}} \sim -\frac{1}{r_{12}^6} \sum_i e_i^{(1)2} r_i^{(1)2} \sum_j e_j^{(2)2} r_j^{(2)2} \frac{1}{\Delta E}, \tag{2.14}$$

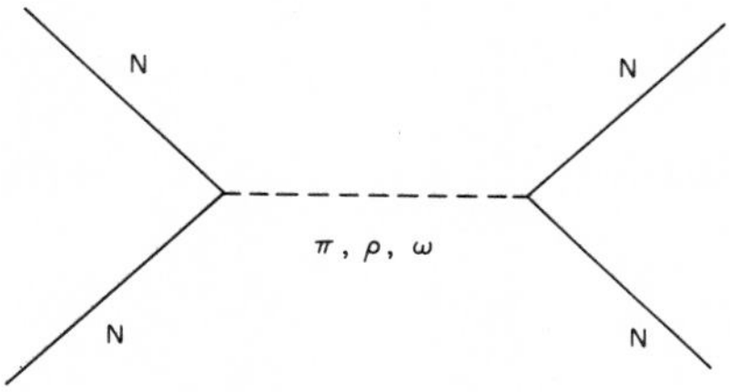

FIGURE 2.3. The basic features of the NN interaction can be understood in terms of a simple exchange of virtual light mesons—π, ρ, ω, etc.—between nucleons.

where ΔE is a typical atomic excitation energy scale. This then is the van der Waals force.

Now let's see if similar insights can be gained in the realm of nuclear physics. The fundamental objects of the nuclear physicist's laboratory, corresponding to the atomic physicist's electron and proton, are, of course, the nucleons—neutron and proton; and the "hydrogen atom," i.e., the simplest bound state, is the deuteron. In a traditional nuclear text one learns that a reasonable semiquantitative picture of the nucleon-nucleon interaction is provided by summing over the exchange of light mesons—π, ρ, ω, η, σ, . . .—between these nucleons, each having its own coupling and Yukawa potential (Figure 2.3) [5],

$$V(r) \sim \sum_i g_i \frac{1}{r} \exp(-m_i r). \tag{2.15}$$

(Here g_i is a generic coupling constant and can contain spin- and isospin-dependent components.) This picture is too simplistic, however, when scrutinized from the perspective of contemporary particle physics. Indeed, the fundamental building blocks, as considered by the particle physicist, are not the nucleons but rather the u,d quarks from which the nucleon is constructed. These quarks are pointlike and quite light objects—$m_{u,d} \sim$ 10 MeV—that possess their own distinctive charge, isospin, etc., quantum numbers, as shown in Table 2.1. The nucleons are three-quark bound states:

$$p\text{: } uud \qquad \begin{aligned} Q &= \left(\frac{2}{3} + \frac{2}{3} - \frac{1}{3}\right)e = e \\ I_z &= \left(\frac{1}{2} + \frac{1}{2} - \frac{1}{2}\right) = \frac{1}{2} \\ B &= \left(\frac{1}{3} + \frac{1}{3} + \frac{1}{3}\right) = 1 \end{aligned} \qquad n\text{: } udd \qquad \begin{aligned} Q &= \left(\frac{2}{3} - \frac{1}{3} - \frac{1}{3}\right)e = 0 \\ I_z &= \left(\frac{1}{2} - \frac{1}{2} - \frac{1}{2}\right) = -\frac{1}{2} \\ B &= \left(\frac{1}{3} + \frac{1}{3} + \frac{1}{3}\right) = 1; \end{aligned}$$

TABLE 2.1
QUANTUM NUMBERS OF THE LIGHT QUARKS

Quark	Charge	I	I_z	B
u	$\frac{2}{3}e$	$\frac{1}{2}$	$\frac{1}{2}$	$\frac{1}{3}$
d	$-\frac{1}{3}e$	$\frac{1}{2}$	$-\frac{1}{2}$	$\frac{1}{3}$
s	$-\frac{1}{3}e$	0	0	$\frac{1}{3}$

Listed above are the quantum numbers of the u, d, s quarks relevant to nuclear physics.

and the mesons are bound states of quark-antiquark pairs:

$$\pi^+: u\bar{d} \qquad Q = \left(\frac{2}{3} - \left(-\frac{1}{3}\right)\right)e = e$$

$$I_z = \frac{1}{2} - \left(-\frac{1}{2}\right) = 1$$

$$B = \frac{1}{3} - \frac{1}{3} = 0.$$

Within such bound systems, quarks interact with one another via exchange of massless quanta termed "gluons" which couple to an additional (and unobserved) "color" quantum number carried both by the quarks *and* by the gluons themselves [6]. The similarity of this picture—in which massless gluons couple to color charge—to quantum electrodynamics (QED)—in which massless photons couple to electric charge—has generated the name quantum chromodynamics (QCD) for this scheme.

Nevertheless, although there are tantalizing similarities between QED and QCD, there exist also important differences that must always be kept in mind. For example, the interquark potential of QED arises from photon exchange and, as shown in the Appendix, is of the form

$$V_{\text{QED}}(r) \sim \frac{\alpha}{r} + \frac{e}{8\pi m r^3} \boldsymbol{\mu} \cdot \mathbf{L} + \boldsymbol{\mu}_1 \times \boldsymbol{\nabla} \cdot \boldsymbol{\mu}_2 \times \boldsymbol{\nabla} \frac{1}{4\pi r} + \ldots. \tag{2.16}$$

However, the QCD potential arises from gluon exchange and is of form

$$V_{\text{QCD}}(r) \sim \frac{\alpha_s}{r} + \frac{\text{“}e_s\text{”}}{8\pi m r^3} \boldsymbol{\mu} \cdot \mathbf{L} + \boldsymbol{\mu}_1 \times \boldsymbol{\nabla} \cdot \boldsymbol{\mu}_2 \times \boldsymbol{\nabla} \frac{1}{4\pi r} + \ldots + \text{“}\lambda r\text{”}. \tag{2.17}$$

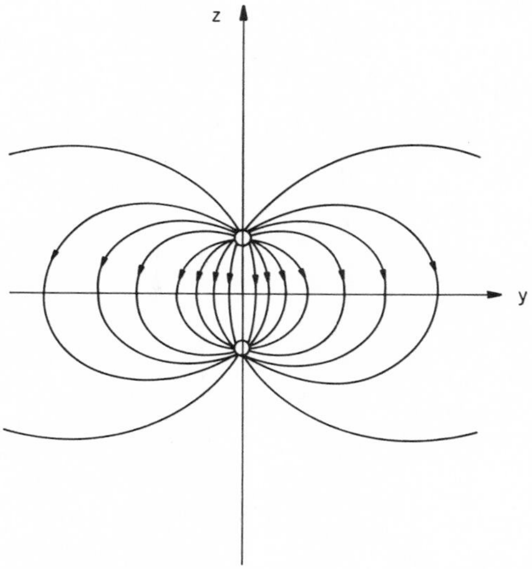

FIGURE 2.4. The electric dipole field in the vicinity of a pair of opposite changes falls off like $1/r^3$ and extends out to distances far beyond the vicinity of the dipole itself.

Here the first three terms are simply color analogs of the corresponding QED potential. (However, the spin-orbit term $\boldsymbol{\mu} \cdot \mathbf{L}$ is generally omitted since ground-state configurations such as the nucleon are purely S-wave so that $\mathbf{L} = 0$.) The key difference between the QCD and QED potentials then is the so-called confinement term λr. Its origin arises from the feature that QCD is a "non-Abelian" theory, whereby the gluons themselves carry color charge and are self-interacting. Thus, whereas an electric dipole in QED has the usual dependence (Figure 2.4) [7]

$$V(r) \sim \frac{\mathbf{p} \cdot \mathbf{r}}{r^3} \tag{2.18}$$

with flux lines extending off to infinity, the $q\bar{q}$ potential in QCD is quite different. Very roughly speaking, because the gluons are self-interacting, the color-flux lines are unable to escape to infinity and form a thin "flux tube" connecting the quark-antiquark pair (Figure 2.5). The energy density per unit length is then roughly constant, giving rise to a linear potential—$V \sim \lambda r$—which is the origin of quark confinement. A precisely linear form is, of course, no doubt simplistic. However, it does give a reasonable semi-quantitative picture of the $q\bar{q}$ interaction at large distances, as shown, for example, in analyses of spectra of heavy quark—$b\bar{b}$, $c\bar{c}$—"atoms." Note

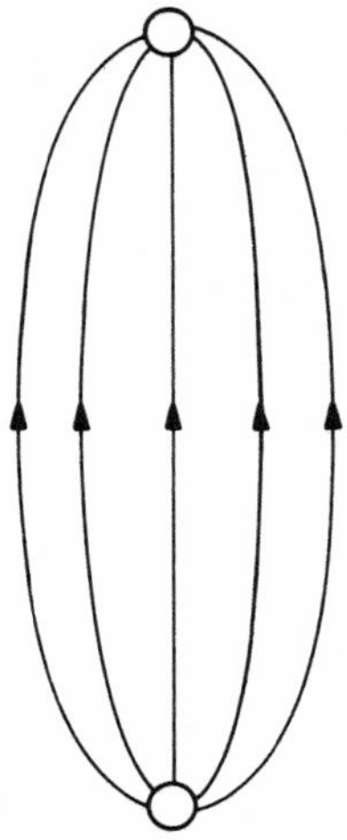

FIGURE 2.5. Due to the self-interaction of gluons, the color-electric fields are confined essentially to the region of space directly between the color-charges.

that "infinity" for large separations in such a potential is actually illusary. One can never separate quarks to infinite distance because beyond certain separation, it becomes energetically favorable instead to produce an additional $q\bar{q}$ pair and to fission into a pair of mesons.

Another curious feature of QCD has to do with the hadronic particle masses. One is familiar with the QED case wherein the attractive (negative) potential between electron and proton generates a hydrogen atom mass which is 13.6 eV *less than* the sum of free electron and proton masses. The analogous QCD situation is quite different, however, since the one GeV nucleon mass is many times *larger than* the ~ 30 MeV sum of quark masses. The explanation is that the binding mechanism is quite different from the familiar atomic physics paradigm. Because of the nonlinear self-couplings of the gluon fields the "vacuum" (i.e., lowest energy state) of QCD in the absence of quark sources is a complex configuration of color fields having energy actually *lower than* the so-called perturbative vacuum, which is the usual null or empty state. However, in the presence of sufficiently strong color fields, such as are generated by the presence of color sources (quarks), there exists a phase transition and the ground state becomes the usual perturbative vacuum. Then in the vicinity of quark sources, a bubble, or "bag," of perturbative vacuum is formed in the vast nonperturbative vacuum sea, and this is what we call a particle [8]. Since the energy of the perturbative vacuum is higher than that of its surroundings, the system attempts to reduce the size of the bubble but is prevented from doing so by the Heisenberg uncertainty principle, which requires that the quarks increase their momentum and hence their pressure on the envelope of the bubble as the radius decreases. The final size of the bubble (particle) results from a balance between these two effects. We shall be more quantitative in this regard later in this section.

One further important difference between QED and QCD involves the relative strength of the hyperfine or dipole-dipole interaction. As already noted, this term plays a relatively small role in the hydrogen atom,

$$\frac{V_2}{V_0} \sim \frac{\alpha}{Mm}\left\langle\frac{1}{r^3}\right\rangle \sim 10^{-7}. \tag{2.19}$$

However, the role of hyperfine splitting in the case of QCD is much more important, as we can see in Table 2.2. In the meson sector the two spin $\frac{1}{2}$ quarks can either be aligned ($s = 1$) or antiparallel ($s = 0$). The resulting particles–ρ and π, respectively—are split by over 600 MeV. In the baryon sector, the total spin of three spin $\frac{1}{2}$ particles can be either $s = \frac{3}{2}$ or $s = \frac{1}{2}$ corresponding to the baryons Δ or N, respectively. The mass splitting here is about 300 MeV. We observe that the hyperfine splitting then is of order 300 MeV or so and plays a key role in particle spectroscopy. We shall return later to the reason for this large hyperfine effect.

The existence of these color degrees of freedom then is crucial to our understanding of hadronic structure. One additional feature is worthy of note here, and that is the role of color in solving the "statistics" problem. Since systems of three (two) quarks are fermions (bosons), respectively, it is clear that the quark itself is a half-integral spin object and therefore should obey the Pauli exclusion principle. This, however, appears to be

TABLE 2.2
PROPERTIES OF LIGHT BARYONS AND MESONS

Meson	Quark Structure	Mass	Spin
π^+	$\bar{d}u$	140 MeV	0
ρ^+	$\bar{d}u$	760 MeV	1
Baryon			
p	uud	940 MeV	$\frac{1}{2}$
Δ^+	uud	1240 MeV	$\frac{3}{2}$

Listed are the masses and spins of the lightest nonstrange baryons and mesons, indicating the effects of the spin-spin coupling.

inconsistent with the existence of the $J = \frac{3}{2}$, $I = \frac{3}{2}$ Δ resonance. Consider the Δ^{++} with $J_z = \frac{3}{2}$, whose quark makeup must be

$$\Delta^{++}\left(s_z = \frac{3}{2}\right): u\uparrow u\uparrow u\uparrow.$$

Here all quarks are assumed to be in relative S-states, so the apparent violation of the Pauli principle is clear, but color comes to the rescue. Assuming the existence of three fundamental color states, the demand that triquark systems such as the nucleon or delta be color neutral requires that such particles be color singlets and totally antisymmetric under color interchange. Then the flavor-spin-space aspect of the wave function is totally symmetric, as found in the case of $\Delta^{++}(s_z = \frac{3}{2})$.

This "evidence" for the presence of color states is, of course, inferential. However, a more direct proof of the validity of the existence of color degrees of freedom is provided by experiments involving the collision of electron and positron beams. We define the ratio of e^+e^- annihilation into hadronic states vs. annihilation into a $\mu^+\mu^-$ pair,

$$R = \frac{\sigma(e^+e^- \rightarrow \text{hadrons})}{\sigma(e^+e^- \rightarrow \mu^+\mu^-)}, \tag{2.20}$$

at a given center of mass energy. To the extent that (1) these processes are the result of the exchange of a single virtual photon between incident and outgoing states, and (2) the strongly interacting final states arise from photon coupling to a virtual quark pair that subsequently fragments completely into hadrons, then, since quarks and muons are both simple pointlike spin $\frac{1}{2}$ systems, we would expect that this ratio R is given simply by

$$R = \sum_i \left(\frac{q_i}{e}\right)^2, \tag{2.21}$$

where q_i,e are the quark, muon charge, respectively, and the sum is over all quark pairs that can be produced at a given center of mass energy. (We assume that we are well above threshold for any given quark pair.) Then in the absence of color degrees of freedom we would expect

$$R^{(0)}_{u,d,s} = \left(\frac{2}{3}\right)^2 + \left(\frac{1}{3}\right)^2 + \left(\frac{1}{3}\right)^2 = \frac{2}{3} \tag{2.22}$$

in the energy range

$$(1\ \text{GeV})^2 < s < (3\ \text{GeV})^2$$

between strange and charm quark thresholds,

$$R^{(0)}_{u,d,s,c} = 2\cdot\left(\frac{2}{3}\right)^2 + 2\cdot\left(\frac{1}{3}\right)^2 = \frac{10}{9} \tag{2.23}$$

in the energy range

$$(3\ \text{GeV})^2 < s < (9\ \text{GeV})^2$$

between charm and b quark thresholds, etc. The existence of N color states would multiply this ratio by the factor N. Comparison with the experimental data, as shown in Figure 2.6, clearly indicates that the ratio R has the behavior expected from the arguments given above and that the data are consistent with the value $N = 3$. This then provides useful quantitative evidence for the existence of color degrees of freedom.

Finally, we note that quark bound states, nucleons, have their own sort of van der Waals force, which is, as in the atomic physics case, completely different in character than the underlying force responsible for the binding. Single quark exchange between nucleons is, of course, forbidden because

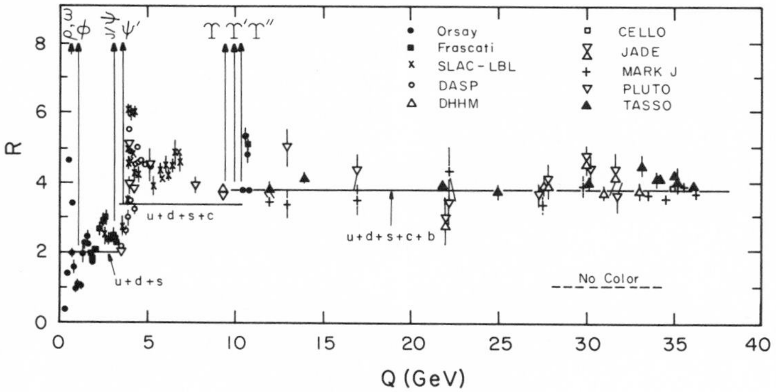

FIGURE 2.6. Plotted is the ratio R of hadron to $\mu^+\mu^-$ production cross sections in e^+e^- annihilation as a function of total e^+e^- center of mass energy. (The sharp spikes correspond to the production of heavy quark-antiquark bound states just below the flavor thresholds.) (*Quarks and Leptons*, F. Halzen and A. Martin, @ 1984, John Wiley & Sons, New York.)

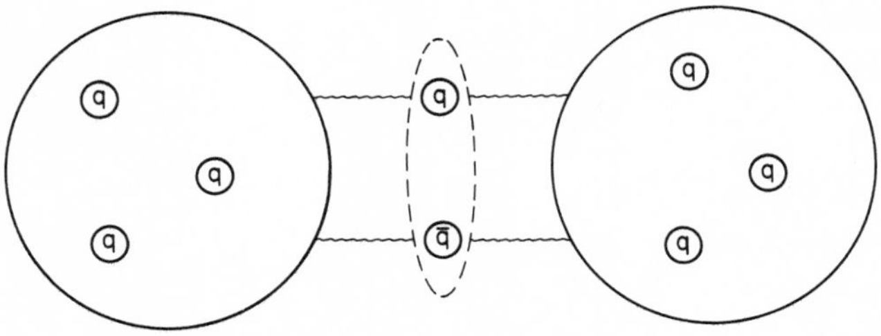

FIGURE 2.7. The Yukawa potential can be considered as due to exchange of a virtual quark-antiquark pair between nucleons.

a single quark cannot exist in a color singlet state. However, exchange of a virtual quark-antiquark pair *is* allowed (cf. Figure 2.7), and this can be considered to represent the usual meson exchange (Yukawa) interaction mentioned above.

Thus far, our arguments have been of a basic handwaving variety. We have seen that a great deal of physics can be explained very successfully at this level in terms of the theory of quark-gluon interactions represented by QCD. If we are willing to be just a bit more quantitative, however, a great deal more can be gleaned, as shown below.

We begin by recalling the Dirac equation for a free particle of mass m, which can be written in terms of a four-component spinor $\psi(x)$ as [9]

$$(i\not{\partial} - m)\psi(x) = 0, \tag{2.24}$$

where

$$\not{\partial} \equiv \gamma^\mu \partial_\mu = \gamma_0 \frac{\partial}{\partial t} + \boldsymbol{\gamma} \cdot \boldsymbol{\nabla}. \tag{2.25}$$

Here we utilize the representation

$$\gamma_0 = \begin{pmatrix} 1 & 0 \\ 0 & -1 \end{pmatrix} \qquad \boldsymbol{\gamma} = \begin{pmatrix} 0 & \boldsymbol{\sigma} \\ -\boldsymbol{\sigma} & 0 \end{pmatrix} \tag{2.26}$$

with

$$\{\gamma_\mu, \gamma_\nu\} = 2g_{\mu\nu} \tag{2.27}$$

for the Dirac matrices. Now, just as Newton's second law,

$$m \frac{d^2\vec{x}}{dt^2} + \vec{\nabla} V(x) = 0, \tag{2.28}$$

is derivable from the Lagrangian

$$L = \frac{1}{2} m\dot{\mathbf{x}}^2 - V(\mathbf{x}) \tag{2.29}$$

via Lagrange's equation of motion

$$\frac{d}{dt}\frac{\partial L}{\partial \dot{\mathbf{x}}} - \frac{\partial L}{\partial \mathbf{x}} = 0, \tag{2.30}$$

then in a corresponding fashion, Dirac's equation

$$(i\not{\partial} - m)\psi(x) = 0 \tag{2.31}$$

is derivable from a Lagrangian density

$$\mathscr{L}(x) = \bar{\psi}(x)(i\not{\partial} - m)\psi(x) \tag{2.32}$$

via the Euler-Lagrange equation of motion

$$\partial_\mu \frac{\delta \mathscr{L}}{\delta(\partial_\mu \bar{\psi})} - \frac{\delta \mathscr{L}}{\delta \bar{\psi}} = 0, \tag{2.33}$$

where $\bar{\psi} \equiv \psi^\dagger \gamma_0$ in standard notation. [Of course, here the "differentiation" is with respect to a function $\bar{\psi}(x)$. However, this can be placed upon a mathematically secure foundation [10], and we need not worry further about this point.]

The piece of the world Lagrangian that is relevant for nuclear physics (i.e., involving only light quarks) then includes the u,d, quark terms

$$\mathscr{L} = \bar{u}(i\not{\partial} - m_u)u + \bar{d}(i\not{\partial} - m_d)d + \dots . \tag{2.34}$$

Here m_u, m_d are small, ~ 10 MeV, but $m_d > m_u$ in order to understand the feature that the neutron is heavier than the proton, even though the proton has more Coulombic energy:

$$\begin{aligned} \Delta m_{\text{Coul}} &= m_{\text{Coul}}(p) - m_{\text{Coul}}(n) = (2q_u q_d + q_u{}^2)\left\langle \frac{1}{4\pi r}\right\rangle \\ &\quad - (2q_u q_d + q_d{}^2)\left\langle \frac{1}{4\pi r}\right\rangle = \frac{\alpha}{3}\left\langle \frac{1}{r}\right\rangle . \end{aligned} \tag{2.35}$$

We can even be quantitative. Taking $R \sim 1$ fm as a typical hadronic size, we find

$$\Delta m_{\text{Coul}} \sim 0.5 \text{ MeV}, \tag{2.36}$$

and thus conclude that, with sizable error bars,

$$m_d - m_u \sim 2 \text{ MeV}. \tag{2.37}$$

The nucleon is then a highly relativistic three-quark structure, since by the Heisenberg uncertainty principle we have

$$\delta p_x \sim \delta p_y \sim \delta p_z \sim \frac{1}{R} \sim 200 \text{ MeV}, \tag{2.38}$$

where again we have taken $R \sim 1$ fm as a typical hadronic scale. Now $\delta p \ggg m$, so

$$E_q \sim p_q \sim \sqrt{\delta p_x{}^2 + \delta p_y{}^2 + \delta p_z{}^2} \sim \frac{\sqrt{3}}{R} \sim 340 \text{ MeV} \tag{2.39}$$

is the energy of a single confined quark, and

$$m_N \sim 3E_q \sim 1 \text{ GeV} \tag{2.40}$$

is the nucleon mass, in good agreement with experiment. Here, $E_q \sim 300$ MeV is often called the "constituent" quark mass m_{cons}, as opposed to the Lagrangian or "current" quark mass m_{curr}. (This distinction between "constituent" and "current" quark mass should always be kept in mind, as it is the origin of a great deal of confusion.)

Since the nucleon is a spin $\frac{1}{2}$ object, it can have a magnetic moment, which can be estimated as follows. In general, the magnetic moment of a system is made up of spin and orbital contributions. The latter is given by

$$\boldsymbol{\mu}_{\text{orbital}} = \int d^3r\mathbf{r} \times \mathbf{J}, \tag{2.41}$$

where

$$\mathbf{J} \sim \rho\mathbf{v} \tag{2.42}$$

is the convective current density. Hence

$$\boldsymbol{\mu}_{\text{orbital}} \sim \frac{e}{2m}\mathbf{L}, \tag{2.43}$$

which then vanishes for systems in the ground state wherein all quarks are in relative S-states. However, since Dirac particles have an intrinsic magnetic moment of size

$$\boldsymbol{\mu}_{\text{spin}} = \frac{e}{2m_{\text{cons}}}\boldsymbol{\sigma}, \tag{2.44}$$

we expect

$$\mu_N \sim \mu_{\text{spin}} \sim \frac{e}{2m_{\text{cons}}} = \frac{3}{2}\frac{e}{m_N}, \tag{2.45}$$

which is the order of magnitude observed experimentally—$\mu_{\text{proton}} = 1.395e/m_N$, $\mu_{\text{neutron}} = -0.955e/m_N$.

We can use this result to estimate the strength of the color spin-spin interaction, for which we predict (cf. eq. [2.9])

$$\begin{aligned} V_{\sigma\cdot\sigma} &\sim \frac{\alpha_{st}}{m_{\text{cons}}^2}\boldsymbol{\sigma}_1\cdot\boldsymbol{\sigma}_2\left\langle\frac{1}{r^3}\right\rangle \\ &\sim \alpha_{st}\frac{1}{m_{\text{cons}}^2 R^3}, \end{aligned} \tag{2.46}$$

where $\alpha_{st} \sim 1$ is the color analog of the electromagnetic fine-structure constant $\alpha_{EM} = \frac{1}{137}$ and is a factor of 10^2 larger since the strong interaction is involved. We anticipate

$$V_{\sigma\cdot\sigma} \sim \frac{\alpha_{st}}{R} \sim 300 \text{ MeV} \tag{2.47}$$

as found experimentally (cf. Table 2.2).

Although we could, in a more extensive discussion, do a great deal more with hadron spectroscopy in this very elementary quark picture, for our purposes it is useful to move on to consider coupling of the quarks to the electromagnetic field described by a vector potential $A_\mu(x)$. In classical electrodynamics this coupling is accomplished by the replacement $p_\mu \to p_\mu - eA_\mu$, which generates the correct interacting relativistic Hamiltonian [11]

$$H = \sqrt{(p - eA)^2 + m^2} \tag{2.48}$$

starting from the simple relativistic energy

$$H = \sqrt{p^2 + m^2}. \tag{2.49}$$

The corresponding quantum mechanical substitution is, of course,

$$i\partial_\mu \to i\partial_\mu - eA_\mu. \tag{2.50}$$

Then the Schrödinger equation for a free charged particle,

$$i\frac{\partial\psi}{\partial t} = -\frac{\nabla^2}{2m}\psi, \tag{2.51}$$

becomes the familiar expression

$$i\frac{\partial\psi}{\partial t} = \left(-\frac{1}{2m}(\nabla + ie\mathbf{A})^2 + e\phi\right)\psi \tag{2.52}$$

when an electromagnetic field is present. Likewise the free Dirac Lagrangian

$$\mathscr{L}_0 = \bar{u}(i\partial\!\!\!/ - m_u)u + \bar{d}(i\partial\!\!\!/ - m_d)d + \ldots \tag{2.53}$$

becomes

$$\begin{aligned} \mathscr{L} &= \bar{u}\left(i\partial\!\!\!/ + \frac{2}{3}eA\!\!\!/ - m_u\right)u + \bar{d}\left(i\partial\!\!\!/ - \frac{1}{3}eA\!\!\!/ - m_d\right)d + \ldots \\ &\equiv \mathscr{L}_0 + eA_\mu J^\mu = \mathscr{L}_0 + \mathscr{L}_{\text{int}} \end{aligned} \tag{2.54}$$

whence we can read off the electromagnetic current,

$$J_\mu = \frac{2}{3}\bar{u}\gamma_\mu u - \frac{1}{3}\bar{d}\gamma_\mu d + \ldots. \tag{2.55}$$

In order to get a feel for the content of this result, consider the plane wave solution to the Dirac equation,

$$\psi(x) = u(p)\exp(-iEt + i\mathbf{p}\cdot\mathbf{x}). \tag{2.56}$$

The form of the plane wave spinor $u(p)$ is found via

$$(i\partial\!\!\!/ - m)\psi(x) = 0 \Rightarrow (p\!\!\!/ - m)u(p)e^{-ip\cdot x} = 0. \tag{2.57}$$

Here

$$p\!\!\!/ - m = E\gamma_0 - \mathbf{p}\cdot\gamma - m = \begin{pmatrix} E - m & -\boldsymbol{\sigma}\cdot\mathbf{p} \\ \boldsymbol{\sigma}\cdot\mathbf{p} & -E - m \end{pmatrix}. \tag{2.58}$$

Writing

$$u(p) = \begin{pmatrix} \beta \\ \lambda \end{pmatrix}, \tag{2.59}$$

where β,λ are both two-component spinors, we find the relations

$$\lambda = \frac{\boldsymbol{\sigma} \cdot \mathbf{p}}{E + m} \beta \qquad \beta = \frac{\boldsymbol{\sigma} \cdot \mathbf{p}}{E - m} \lambda, \tag{2.60}$$

which are consistent with one another provided

$$\boldsymbol{\sigma} \cdot \mathbf{p} \boldsymbol{\sigma} \cdot \mathbf{p} = \mathbf{p}^2 = E^2 - m^2, \tag{2.61}$$

which is the usual relationship between relativistic energy and momentum. Now define

$$\beta \equiv N\chi, \tag{2.62}$$

where $\chi^\dagger \chi = 1$, and choose N by the condition

$$\bar{u}(p)u(p) = 1. \tag{2.63}$$

This yields

$$u(p) = N \begin{pmatrix} \chi \\ \dfrac{\boldsymbol{\sigma} \cdot \mathbf{p}}{E + m} \chi \end{pmatrix}, \tag{2.64}$$

where $N = \sqrt{E + m/2m}$ and χ is the usual two-component (Pauli) spinor. In the nonrelativistic limit—$E \approx m$, $p/m \approx v/c$—we find

$$\begin{aligned} J_0 &\sim \bar{u}\gamma_0 u = u^\dagger u \sim \chi^\dagger \chi \left(1 + \mathcal{O}\left(\frac{v^2}{c^2}\right)\right) \\ \mathbf{J} &\sim \bar{u}\gamma u = u^\dagger \gamma_0 \gamma u \sim \chi^\dagger \chi \frac{\mathbf{p}}{m} + \ldots = J_0 \mathbf{v}, \end{aligned} \tag{2.65}$$

which have the expected forms.

It is also interesting to examine the behavior of the current J_μ under a "spatial inversion" or parity transformation,

$$\begin{aligned} \mathbf{r} &\xrightarrow[P]{} -\mathbf{r} \\ t &\xrightarrow[P]{} t. \end{aligned} \tag{2.66}$$

Clearly, then

$$\mathbf{v} = \frac{d\mathbf{r}}{dt} \xrightarrow[P]{} -\frac{d\mathbf{r}}{dt} = -\mathbf{v}$$
$$\mathbf{L} = \mathbf{r} \times m\mathbf{v} \xrightarrow[P]{} -\mathbf{r} \times m(-\mathbf{v}) = +\mathbf{L}. \tag{2.67}$$

Since spin and orbital angular momentum must transform in the same way (after all, $\mathbf{J} = \mathbf{L} + \mathbf{S}$ is a good quantum number), we also require

$$\mathbf{S} \xrightarrow[P]{} \mathbf{S}. \tag{2.68}$$

In order to study the transformation properties of the electromagnetic current, we need to know the effect of a parity transformation on the Dirac solution $\psi(x)$,

$$\psi(\mathbf{x},t) \xrightarrow[P]{} \gamma_0\psi(-\mathbf{x},t). \tag{2.69}$$

That this is the correct form is clear, since

$$\gamma_0 u(p,s) = N\begin{pmatrix}1 & 0\\ 0 & -1\end{pmatrix}\begin{pmatrix}\chi_s \\ \dfrac{\boldsymbol{\sigma}\cdot\mathbf{p}}{E+m}\chi_s\end{pmatrix} = N\begin{pmatrix}\chi_s \\ \dfrac{\boldsymbol{\sigma}\cdot(-\mathbf{p})}{E+m}\chi_s\end{pmatrix} = u(-\mathbf{p},s) \tag{2.70}$$

and

$$e^{-iEt+i\mathbf{p}\cdot(-\mathbf{x})} = e^{-iEt+i(-\mathbf{p})\cdot\mathbf{x}}. \tag{2.71}$$

The current J_μ thus transforms under spatial inversion as

$$\begin{aligned} J_0 &\xrightarrow[P]{} \psi^\dagger\gamma_0\gamma_0\psi = \bar{\psi}\gamma_0\psi = J_0 & \rho &\to \rho \\ \mathbf{J} &\xrightarrow[P]{} \psi^\dagger\gamma_0\gamma_0\boldsymbol{\gamma}\gamma_0\psi = -\bar{\psi}\boldsymbol{\gamma}\psi = -\mathbf{J} & \rho\mathbf{v} &\to -\rho\mathbf{v}, \end{aligned} \tag{2.72}$$

which is the signature of a "polar vector."

In order to discuss the weak interactions, we also need to construct an "axial vector," which has the opposite transformation properties. For this purpose, we define

$$\gamma_5 = -i\gamma_0\gamma_1\gamma_2\gamma_3 = \begin{pmatrix}0 & -1\\ -1 & 0\end{pmatrix} \tag{2.73}$$

and

$$A_\mu \equiv \bar{\psi}\gamma_\mu\gamma_5\psi. \tag{2.74}$$

Then, since

$$\{\gamma_5,\gamma_\mu\} = 0, \tag{2.75}$$

we find

$$\begin{aligned} A_0 &\xrightarrow[P]{} \bar{\psi}\gamma_0\gamma_0\gamma_5\gamma_0\psi = -\bar{\psi}\gamma_0\gamma_5\psi = -A_0 \\ \mathbf{A} &\xrightarrow[P]{} \bar{\psi}\gamma_0\boldsymbol{\gamma}\gamma_5\gamma_0\psi = +\bar{\psi}\boldsymbol{\gamma}\gamma_5\psi = +\mathbf{A}, \end{aligned} \tag{2.76}$$

which is the characteristic transformation property of an axial vector. As we shall detail later, the weak interaction is maximally parity violating in that an equal mixture of polar and axial currents is involved:

$$J_\mu{}^W = V_\mu + A_\mu = \bar{\psi}\gamma_\mu(1 + \gamma_5)\psi. \tag{2.77}$$

Another way of looking at spatial inversion is in terms of "handedness." Assuming conservation of angular momentum, parity is equivalent to reflection in a mirror, since reflection plus rotation by 180° about an axis normal to the mirror gives

$$\begin{aligned} \text{Ref}(z) \otimes \text{Rot}_z(\pi) &= (x,y,-z) \otimes (-x,-y,z) \\ &= (-x,-y,-z) = P. \end{aligned} \tag{2.78}$$

However, when a right (left) hand is reflected in a mirror, it becomes a left (right) hand. Thus, another way of saying that the world is parity invariant is to require the absence of physics that would select a particular handedness. Feynman refers to this as the inability to construct an experiment that would enable some extraterrestrial civilization with whom we are in radio contact to distinguish the concept of left/right [12]. Until the mid-1950s essentially all physicists believed this handedness symmetry to be valid, and they were stunned when experiment indicated that the weak force selected out a particular handedness.

This aspect of the weak interaction can best be discussed by looking at "helicity" eigenstates of the Dirac equation. A negative helicity or left-handed spinor is one whose spin and momentum lie in opposite directions and satisfies

$$\overset{\hat{s}}{\Leftarrow}\!\!\xrightarrow[\hat{p}]{} \quad \boldsymbol{\sigma}\cdot\hat{p}\chi_L = -\chi_L, \tag{2.79}$$

while a positive helicity or right-handed spinor is one whose spin and momentum are parallel and obeys

$$\overset{\hat{s}}{\Rightarrow}\!\!\xrightarrow[\hat{p}]{} \quad \boldsymbol{\sigma}\cdot\hat{p}\chi_R = +\chi_R. \tag{2.80}$$

Now, the Dirac spinor $u(p)$ becomes in the high energy limit $E \gg m$ or for massless particles (such as neutrinos)

$$u(p) \sim N' \begin{pmatrix} \chi \\ \boldsymbol{\sigma} \cdot \hat{p}\chi \end{pmatrix}, \tag{2.81}$$

so that a left- or right-handed Dirac particle has the simple representation

$$u_L = N' \begin{pmatrix} \chi \\ -\chi \end{pmatrix} \qquad u_R = N' \begin{pmatrix} \chi \\ \chi \end{pmatrix}. \tag{2.82}$$

In this limit, the so-called chirality operators

$$\frac{1}{2}(1 \pm \gamma_5) \tag{2.83}$$

also act as projection operators for left- and right-handed helicity states:

$$\begin{aligned} \frac{1}{2}(1 + \gamma_5)u_L &= u_L \qquad & \frac{1}{2}(1 + \gamma_5)u_R &= 0 \\ \frac{1}{2}(1 - \gamma_5)u_L &= 0 \qquad & \frac{1}{2}(1 - \gamma_5)u_R &= u_R. \end{aligned} \tag{2.84}$$

In terms of chirality eigenstates, the weak current assumes the simple form

$$J_\mu{}^W = 2\bar{\psi}_L \gamma_\mu \psi_L, \tag{2.85}$$

where

$$\psi_L \equiv \frac{1}{2}(1 + \gamma_5)\psi. \tag{2.86}$$

It is in this sense that the weak current has maximal "handedness" and thus violates parity maximally.

The form of the effective weak interaction at low energies was written down by Fermi over fifty years ago [13]

$$H_W = \frac{G}{\sqrt{2}} J_\mu{}^W \otimes J_W^{\mu\dagger}, \tag{2.87}$$

where $G \approx 10^{-5} m_p^{-2}$ is a universal coupling constant and the weak current $J_\mu{}^W$ has the form

$$\frac{1}{2} J_\mu{}^W = \bar{\nu}_{\mu L} \gamma_\mu \mu_L + \bar{\nu}_{eL} \gamma_\mu e_L + \bar{\nu}_{\tau L} \gamma_\mu \tau_L$$
$$+ \cos \theta_c \bar{u}_L \gamma_\mu d_L + \sin \theta_c \bar{u}_L \gamma_\mu s_L + \ldots . \quad (2.88)$$

Here $\theta_c \approx 15°$ is the so-called Cabibbo angle and measures the relative strengths of strangeness-changing and strangeness-preserving couplings [14].

We predict, for example, that in a nuclear beta decay reaction, involving

$$H_\beta = \frac{4G}{\sqrt{2}} \cos \theta_c \bar{u}_L \gamma_\mu d_L \otimes \bar{e}_L \gamma^\mu \nu_{eL}, \quad (2.89)$$

the emitted anti-neutrino (assumed to be massless) should be purely right handed while the electron should be dominantly left handed (at least to the extent that $E_e \gg m_e$). Both of these predictions are well borne out experimentally. For example, direct measurements of electron/positron helicity have given [15]

$$P_L^{\text{expt}} = (0.99 \pm 0.01) \frac{v}{c}, \quad (2.90)$$

to be compared with the value

$$P_L^{\text{theo}} = \frac{v}{c} \quad (2.91)$$

expected from eq. (2.89). Experiments involving neutrinos are much less precise because of the very tiny cross sections involved. However, here too experiment is consistent with the simple left-handed form [16].

Now, those somewhat familiar with this material might well ask: But what about neutron beta decay? We know in this case that

$$\langle p | J_\mu{}^W | n \rangle = \bar{u}_p [\gamma_\mu (g_V + g_A \gamma_5) + \ldots] u_n \quad (2.92)$$

with

$$g_V(q^2 = 0) = 1 \qquad g_A(q^2 = 0) = 1.25. \quad (2.93)$$

This result appears inconsistent with the simple, purely left-handed form of the current postulated in eq. (2.77). This effect, however, is simply a renormalization due to the fact that the neutron/proton are bound states of the fundamental objects of which they are composed—quarks. Thus, we have

$$\begin{aligned} |p\uparrow\rangle &= \frac{1}{\sqrt{18}}\{uud(\uparrow\downarrow\uparrow + \downarrow\uparrow\uparrow - 2\uparrow\uparrow\downarrow) \\ &\quad + udu(\uparrow\uparrow\downarrow + \downarrow\uparrow\uparrow - 2\uparrow\downarrow\uparrow) + duu(\uparrow\downarrow\uparrow + \uparrow\uparrow\downarrow - 2\downarrow\uparrow\uparrow)\} \qquad (2.94) \\ |n\uparrow\rangle &= \text{Same with } (u \leftrightarrow d). \end{aligned}$$

If we now evaluate the relevant matrix elements, we observe that the polar vector current normalization is unchanged (which is essential so that there is no renormalization of the electric charge). Thus, in the nonrelativistic limit

$$\begin{gathered} V_\mu\colon V_0 \sim \delta_{ss'}u^\dagger d \\ \langle p\uparrow|V_0|n\uparrow\rangle = 3\cdot\left(\frac{1}{\sqrt{18}}\right)^2\cdot(8-2) = 1. \end{gathered} \qquad (2.95)$$

However, for the axial current this is *not* the case, and there is a strong renormalization. In the nonrelativistic limit

$$\begin{gathered} A_\mu\colon \mathbf{A} \sim \boldsymbol{\sigma}_{s's}u^\dagger d \\ \langle p\uparrow|A_3|n\uparrow\rangle = 3\cdot\left(\frac{1}{\sqrt{18}}\right)^2\cdot(8+2) = \frac{5}{3}. \end{gathered} \qquad (2.96)$$

Here $\frac{5}{3} = 1.667$ is quite large compared to the value 1.25 seen experimentally. Our simple quark picture then appears to fail. Not so. Remember that u,d quarks are very light—~ 10 MeV—so that a nonrelativistic approach is unwarranted. Rather, a proper relativistic wave function,

$$\psi(r) = \begin{pmatrix} u(r)\chi_s \\ \boldsymbol{\sigma}\cdot\hat{r}l(r)\chi_s \end{pmatrix}, \qquad (2.97)$$

which includes both upper *and* lower components, is required. One must, of course, normalize the wave function to unity,

$$\int d^3r(u^2(r) + l^2(r)) = 1. \qquad (2.98)$$

Matrix elements of the polar and axial currents can now be taken, yielding

$$\langle p\uparrow|V_0|n\uparrow\rangle = 1 \times \int d^3r(u^2(r) + l^2(r)) = 1$$
$$\langle p\uparrow|A_3|n\uparrow\rangle = \frac{5}{3} \times \int d^3r\left(u^2(r) - \frac{1}{3}l^2(r)\right) \sim 1.1 - 1.3, \tag{2.99}$$

where the specific value depends on the form of the quark wave function being employed and the assumed masses for the quarks. For example, in the MIT bag model with massless quarks, we have [17]

$$g_A/g_V = 1.09. \tag{2.100}$$

Such "relativistic" corrections are then not optional—they are essential.

We can now understand many basic features of hadron spectroscopy and weak interaction phenomenology from the point of view of simple quark ideas. However, rather than continue along this line, we will switch gears and examine in the next chapter the motivation for the fundamental weak interaction itself.

Appendix

We here consider the effective potential between electron and proton which arises from the idea of the e–p interaction as being due to the exchange of a virtual photon between these two systems. We begin with the result from electrodynamics that the action describing the interaction of the electromagnetic field and a charged particle is given by

$$S = e_1 \int d^4x J_\mu^{(1)}(x) A^\mu(x), \tag{A1}$$

where $A^\mu(x)$ is the vector potential and $e_1 J_\mu^{(1)}(x)$ is the (four-) current density associated with the particle, hereafter called #1. If the vector potential is produced by the presence of a second charged particle, #2, then the Maxwell equations require that

$$\Box_x A_\mu(x) = e_2 J_\mu^{(2)}(x), \tag{A2}$$

whose solution

$$A^\mu(x) = i \int d^4y D^{\mu\nu}(x-y) e_2 J_\nu^{(2)}(y) \tag{A3}$$

is given in terms of the Green's function,

$$D^{\mu\nu}(x - y) = \int \frac{d^4q}{(2\pi)^4} e^{-iq\cdot(x-y)} \frac{-i}{q^2 + i\varepsilon} g^{\mu\nu}. \tag{A4}$$

The action for this interacting system then becomes

$$S = ie_1e_2 \int d^4x \int d^4y J_\mu^{(1)}(x) D^{\mu\nu}(x - y) J_\nu^{(2)}(y), \tag{A5}$$

which is also the lowest-order transition amplitude according to quantum mechanics when $J_\mu(x)$ are given in terms of their proper quantum mechanical representations.

Recall that the Lagrangian for a free Dirac particle of mass m is given by

$$\mathscr{L}_0(x) = \bar{\psi}(i\gamma_\mu\partial^\mu - m)\psi. \tag{A6}$$

Interaction with the electromagnetic field is described by the well-known classical substitution

$$p_\mu \to p_\mu - eA_\mu, \tag{A7}$$

which has the quantum mechanical analog

$$i\partial_\mu \to i\partial_\mu - eA_\mu. \tag{A8}$$

The Lagrangian then becomes

$$\begin{aligned} \mathscr{L}(x) &= \bar{\psi}(x)(i\partial\!\!\!/ - eA\!\!\!/(x) - m)\psi(x) \\ &\equiv \mathscr{L}_0(x) + \mathscr{L}_{\text{int}}(x). \end{aligned} \tag{A9}$$

Writing

$$\mathscr{L}_{\text{int}}(x) \equiv eJ_\mu(x)A^\mu(x), \tag{A10}$$

we recognize

$$J_\mu(x) = \bar{\psi}(x)\gamma_\mu\psi(x) \tag{A11}$$

as the conserved probability current density that we require.

Thus for a process wherein particle #1 of mass m, charge e_1 and particle #2 of mass M, charge e_2 make a transition from initial momentum states p_1,P_1 to outgoing momentum states p_2,P_2, respectively, the lowest-order quantum mechanical amplitude becomes

$$\begin{aligned} S &= i \int d^4x \int d^4y \int \frac{d^4q}{(2\pi)^4} e^{i(p_2 - p_1 - q)\cdot x} e^{i(P_2 - P_1 + q)\cdot y} \\ &\quad \times \frac{mM}{\sqrt{\varepsilon_1 \varepsilon_2 E_1 E_2}} e_1 \bar{u}(p_2)\gamma_\mu u(p_1) \frac{-ig^{\mu\nu}}{q^2 + i\varepsilon} e_2 \bar{u}(P_2)\gamma_\nu u(P_1) \\ &= (2\pi)^4 \delta^4(p_2 + P_2 - p_1 - P_1) \frac{mM}{\sqrt{\varepsilon_1 \varepsilon_2 E_1 E_2}} e_1 e_2 \bar{u}(p_2)\gamma_\mu u(p_1) \\ &\quad \times \frac{1}{(p_2 - p_1)^2 + i\varepsilon} \bar{u}(P_2)\gamma^\mu u(P_1). \end{aligned} \tag{A12}$$

In the nonrelativistic limit $\frac{p_1}{m} \ll 1$, $\frac{P_1}{M} \ll 1$ the Dirac spinors assume the simple form

$$u(p) \approx \left(1 + \frac{p^2}{8m^2}\right)\begin{pmatrix} \chi \\ \frac{\boldsymbol{\sigma}\cdot\mathbf{p}}{2m}\chi \end{pmatrix}, \tag{A13}$$

and the $e - p$ action becomes

$$\begin{aligned} S &\approx (2\pi)^4 \delta^4(p_2 + P_2 - p_1 - P_1) e^2 \left(1 - \frac{p_1{}^2 + p_2{}^2}{8m^2}\right) \frac{-1}{(\mathbf{p}_1 - \mathbf{p}_2)^2} \\ &\quad \times \Bigg[\tilde{\chi}_2{}^\dagger \tilde{\chi}_1 \chi_2{}^\dagger \left(1 + \frac{1}{4m^2}\mathbf{p}_2 \cdot \mathbf{p}_1 + \frac{i}{4m^2}\boldsymbol{\sigma}\cdot(\mathbf{p}_2 \times \mathbf{p}_1)\right)\chi_1 \\ &\quad - \tilde{\chi}_2{}^\dagger \left(\frac{\mathbf{P}_1 + \mathbf{P}_2}{2M} - \frac{i}{2M}\boldsymbol{\sigma} \times (\mathbf{P}_1 - \mathbf{P}_2)\right)\tilde{\chi}_1 \cdot \chi_2{}^\dagger \\ &\quad \times \left(\frac{\mathbf{p}_1 + \mathbf{p}_2}{2m} - \frac{i}{2m}\boldsymbol{\sigma} \times (\mathbf{p}_1 - \mathbf{p}_2)\right)\chi_1 \Bigg], \end{aligned} \tag{A14}$$

where the $\sim$ indicates a proton spinor.

We may now identify the various interaction terms by recalling that the Born approximation requires the transition amplitude and the corresponding potential to be Fourier transforms of one another:

$$\begin{aligned} S_{fi} &= \langle f|V|i\rangle = \int d^3r e^{-i(\mathbf{p}_f - \mathbf{p}_i)\cdot\mathbf{r}} V(\mathbf{r}) \\ V(r) &= \int \frac{d^3q}{(2\pi)^3} e^{-i\mathbf{q}\cdot\mathbf{r}} S_{fi}. \end{aligned} \tag{A15}$$

Recalling that

$$\int \frac{d^3q}{(2\pi)^3} e^{-i\mathbf{q}\cdot\mathbf{r}} \frac{1}{\mathbf{q}^2} = \frac{1}{2\pi^2} \int_0^\infty dq\, j_0(qr) = \frac{1}{4\pi r}, \tag{A16}$$

we recognize the leading term (independent of recoil)

$$V_0(r) = -\frac{e^2}{4\pi r} \tilde{\chi}_2{}^\dagger \tilde{\chi}_1 \chi_2{}^\dagger \chi_1 \tag{A17}$$

as simply the dominant Coulomb interaction between electron and proton.

A second component is also easily identified. For a spherically symmetric potential $V_0(r)$ we have

$$\int d^3r e^{-i\mathbf{p}_2\cdot\mathbf{r}} \frac{1}{r}\frac{dV_0}{dr} \boldsymbol{\sigma}\cdot\mathbf{r} \times (-i\boldsymbol{\nabla}) e^{i\mathbf{p}_1\cdot\mathbf{r}} = -i\boldsymbol{\sigma}\cdot\mathbf{p}_1 \times \mathbf{p}_2 \int d^3r e^{i\mathbf{q}\cdot\mathbf{r}} V_0(r) \tag{A18}$$

so that the spin-dependent piece in the first line of eq. (A14) corresponds to the interaction potential

$$\frac{1}{4m^2}\frac{1}{r}\frac{dV_0}{dr} \chi_2{}^\dagger \boldsymbol{\sigma}\cdot\mathbf{L}\chi_1 \tilde{\chi}_2{}^\dagger \tilde{\chi}_1, \tag{A19}$$

which is the usual spin-orbit term. (Note that the Thomas precession is already included here.)

Finally, including the remaining $\mathbf{p}_1, \mathbf{p}_2$ terms from the first line of eq. (A14) we cancel the $\mathbf{q}^2$ term in the denominator, yielding

$$-\frac{e^2}{8m^2} \chi_2{}^\dagger \chi_1 \tilde{\chi}_2{}^\dagger \tilde{\chi}_1 = \int d^3r e^{-i\mathbf{q}\cdot\mathbf{r}} \left(-\delta^3(r) \frac{e^2}{8m^2} \chi_2{}^\dagger \chi_1 \tilde{\chi}_2{}^\dagger \tilde{\chi}_1 \right), \tag{A20}$$

which is just the Darwin term, since we recognize

$$e\delta^3(r) = -\frac{e}{4\pi} \nabla^2 \frac{1}{r} = \boldsymbol{\nabla}\cdot\mathbf{E}. \tag{A21}$$

Having thus identified each piece of the top line of eq. (A14) with a familiar component of the effective nonrelativistic interaction Hamiltonian, we move now to the terms on the second line of the photon exchange interaction. Since

$$\frac{\mathbf{p}_1 + \mathbf{p}_2}{2m}$$

represents the component of the current density due to motion (i.e., convection), while

$$-\frac{i}{2m}\boldsymbol{\sigma} \times (\mathbf{p}_1 - \mathbf{p}_2)$$

designates the contribution from the intrinsic magnetic moment, we observe that this additional interaction energy consists of three pieces: (1) a current-current interaction; (2) a current-dipole interaction; and (3) a dipole-dipole interaction. Such terms are expected from classical-physics considerations and were written down by Breit before being derived quantum mechanically. For this reason they are often referred to as the Breit interaction.

For example, we note that the dipole-dipole interaction component is equivalent to the potential

$$\begin{aligned} V_{\text{dipole-dipole}} &= \int \frac{d^3q}{(2\pi)^3}\frac{e^2}{\mathbf{q}^2}\chi_2{}^\dagger\boldsymbol{\sigma}\chi_1 \times \mathbf{q}\cdot\tilde{\chi}_2{}^\dagger\boldsymbol{\sigma}\tilde{\chi}_1 \times \mathbf{q}e^{i\mathbf{q}\cdot\mathbf{r}}\frac{1}{2m\cdot 2M} \\ &= -\frac{e}{2m}\chi_2{}^\dagger\boldsymbol{\sigma}\chi_1\cdot\nabla\times\left(\frac{e}{2M}\tilde{\chi}_2{}^\dagger\boldsymbol{\sigma}\tilde{\chi}_1\times\nabla\frac{1}{4\pi r}\right). \end{aligned} \tag{A22}$$

However, we recognize

$$\mathbf{B}(\mathbf{r}) = \nabla \times \left(\frac{e}{2M}\tilde{\chi}_2{}^\dagger\boldsymbol{\sigma}\tilde{\chi}_1 \times \nabla\frac{1}{4\pi r}\right) \tag{A23}$$

as being the magnetic field produced by a proton magnetic dipole moment (for an "ideal" proton with $g_p = 2$). The dipole-dipole interaction can then be represented as

$$V_{\text{dipole-dipole}} = -\boldsymbol{\mu}_e \cdot \mathbf{B}, \tag{A24}$$

which is the energy of the magnetic dipole moment of the "ideal" electron arising from the presence of the magnetic field produced by the ("ideal") proton moment. Such a potential is well known, of course, and is termed the "hyperfine" interaction. In the hydrogen atom the ground state is split by the hyperfine potential into a component with $F = 1$ and one with

$F = 0$, where

$$\mathbf{F} = \mathbf{S}_e + \mathbf{S}_p \tag{A25}$$

is the total atomic angular momentum. (The term "hyperfine" is used because, while the splitting is the same order in α as that found for fine structure, there is an additional suppression here in the amount m/M because of the smallness of the proton magneton.) The value of this splitting

$$\nu = \frac{\omega}{2\pi} = 1.42040575180(3) \times 10^9 \text{ Hz} \tag{A26}$$

is one of the most precisely measured constants of nature and is responsible for the famous 21-cm radiation observed by radioastronomers.

For completeness it is necessary to note that the realistic electron-proton interaction requires two modifications. The first is that experimentally the g-factor of the proton is not $g_p = 2$, as would be the case for an "ideal" Dirac particle, but is rather

$$g_p^{\text{exp}} = 5.58, \tag{A27}$$

which should be used whenever the proton moment is required. This shift in the gyromagnetic ratio is associated with the feature that the proton, unlike the electron, is *not* simply a point particle but has a complex structure in terms of quarks, mesons, etc.

The second modification is also associated with this nonpointlike nature. Describing the proton in terms of a charge *distribution* $\rho(\mathbf{r})$, the Coulomb potential is not given simply by α/r but rather by

$$V(r) = \int d^3s \frac{\alpha}{|\mathbf{r} - \mathbf{s}|} \rho(\mathbf{s}) \tag{A28}$$

(where we assume the normalization $\int d^3s\rho(s) = 1$). Because of this "finite proton size" the lowest-order scattering amplitude becomes

$$\begin{aligned} S_{fi} &= \int d^3r e^{i\mathbf{q}\cdot\mathbf{r}} V(r) = \int d^3r \int d^3s e^{i\mathbf{q}\cdot\mathbf{r}} \frac{\alpha}{|\mathbf{r} - \mathbf{s}|} \rho(\mathbf{s}) \\ &= \int d^3s e^{i\mathbf{q}\cdot\mathbf{s}} \rho(\mathbf{s}) \int d^3w \frac{\alpha}{|\mathbf{w}|} e^{i\mathbf{q}\cdot\mathbf{w}}. \end{aligned} \tag{A29}$$

Defining

$$F(\mathbf{q}) = \int d^3s e^{i\mathbf{q}\cdot\mathbf{s}} \rho(\mathbf{s}), \tag{A30}$$

we see that the finite size effect is given simply by multiplying the point scattering amplitude by this "form factor" $F(\mathbf{q})$,

$$S_{fi} = F(\mathbf{q}) \times S_{fi}^{\text{point-charge}}. \tag{A31}$$

The resulting cross section becomes

$$\frac{d\sigma}{d\Omega} = \left(\frac{d\sigma}{d\Omega}\right)^{\text{point}} |F(\mathbf{q})|^2. \tag{A32}$$

Here

$$\mathbf{q}^2 = 4p_1{}^2 \sin^2 \frac{\theta}{2}, \tag{A33}$$

and thus by careful measurement of the cross section the form factor $F(\mathbf{q})$ (or at least its norm) can be evaluated. The associated charge density $\rho(\mathbf{r})$ can then be determined by taking the inverse Fourier transform

$$\rho(\mathbf{r}) = \int \frac{d^3q}{(2\pi)^3}\, e^{-i\mathbf{q}\cdot\mathbf{r}} F(\mathbf{q}). \tag{A34}$$

Such programs have resulted in a very precise mapping of the proton and nuclear charge distributions.

We have seen then how the picture of the electromagnetic interaction in terms of virtual photon exchange includes automatically all expected features of the e–p potential and offers a concise and correct representation of this interaction.

References

[1] See, e.g., E. Merzbacher, *Quantum Mechanics*, Wiley, New York (1961), Ch. 17.
[2] R. P. Feynman, *Quantum Electrodynamics*, Benjamin, New York (1962), Ch. 3.
[3] J. D. Bjorken and S. D. Drell, *Relativistic Quantum Mechanics*, McGraw-Hill, New York (1964), Ch. 4.
[4] C. Itzykson and J. Zuber, *Quantum Field Theory*, McGraw-Hill, New York (1980), sec. 3-2-4.
[5] M. M. Nagels et al., Phys. Rev. *D12*, 744 (1975), and *D15*, 2547 (1977).
[6] An elementary discussion is provided in G. Kane, *Modern Elementary Particle Physics*, Addison-Wesley, Reading, Mass. (1987),

Ch. 7. See also F. Halzen and A. Martin, *Quarks and Leptons*, Wiley, New York (1984), Ch. 10.

[7] See, e.g., D. J. Griffiths, *Introduction to Electrodynamics*, Prentice-Hall, Englewood Cliffs, N.J. (1981), Ch. 3.

[8] A. Chodos, R. L. Jaffe, K. Johnson, C. B. Thorn, and V. F. Weisskopf, Phys. Rev. *D9*, 3471 (1974).

[9] J. D. Bjorken and S. D. Drell, ref. 3, Ch. 1.

[10] J. D. Bjorken and S. D. Drell, *Relativistic Quantum Fields*, McGraw-Hill, New York (1964), Ch. 11.

[11] The validity of the substitution $p - eA$ in classical physics is demonstrated, e.g., in A. L. Fetter and J. D. Walecka, *Theoretical Mechanics of Particles and Continua*, McGraw-Hill, New York (1980), Ch. 6.

[12] R. P. Feynman, R. B. Leighton, and M. B. Sands, *The Feynman Lectures on Physics*, Addison-Wesley, Reading, Mass. (1963), Vol. I, Ch. 53.

[13] E. Fermi, Z. Phys. *88*, 161 (1934); see translation by A. L. Wilson, Am. J. Phys. *36*, 1150 (1968).

[14] N. Cabibbo, Phys. Rev. Lett. *10*, 531 (1963).

[15] See, e.g., A. R. Brosi et al., Nucl. Phys. *33*, 353 (1962); F.W.J. Koks and J. Van Klinken, Nucl. Phys. *A272*, 61 (1976); J. D. Ullman et al., Phys. Rev. *122*, 536 (1961).

[16] M. Goldhaber, L. Grodzins, and A. W. Sunyar, Phys. Rev. *109*, 1015 (1958); J. C. Palathingal, Bull. Am. Phys. Soc. *14*, 587 (1969).

[17] T. DeGrand, R. Jaffe, K. Johnson, and J. Kiskis, Phys. Rev. *D12*, 2060 (1975).

Chapter 3

THE WEINBERG-SALAM MODEL

. . . the unparalleled embarrassment of a harassed pedlar gauging the symmetry of her peeled pears.

—JAMES JOYCE, *ULYSSES*

Every physicist now knows that the electroweak interactions are well described by the so-called Weinberg-Salam model [1]. The derivation of this interaction is so elegant and simple that it is worth doing at least once, just for the fun of seeing how straightforwardly things work out. We will do so in this chapter.

We begin with a brief digression—noting that the Dirac Lagrangian for a freely moving particle of mass m is invariant under a change in phase of the wavefunction; that is,

$$\mathscr{L}(x) = \bar{\psi}(x)(i\not{\partial} - m)\psi(x) \rightarrow \mathscr{L}(x) \tag{3.1}$$

under

$$\begin{aligned} \psi(x) &\rightarrow e^{i\chi}\psi(x) \\ \bar{\psi}(x) &\rightarrow e^{-i\chi}\bar{\psi}(x) \end{aligned} \tag{3.2}$$

if χ is taken as a constant. In the parlance of mathematicians this is termed a "global U(1) invariance." It is called U(1) since it is unitary—$e^{-i\chi}e^{i\chi} = 1$—and has a single parameter χ. It is global in the sense that the phase change is identical at all points in spacetime.

However, this requirement of global invariance seems rather restrictive. Why must the phase change here on earth be the same as that behind the moon or near Cygnus X-3? Shouldn't there exist rather a theory with a "local U(1) symmetry," wherein the phase χ can depend on x? The change in phase at *our* location need have nothing to do with components of the wave function in *other* parts of the universe. We can construct such a locally U(1) invariant theory if we are willing to pay a price: introduction of an extra "gauge field," $A_\mu(x)$. Defining the "covariant derivative" (cf. eq. [2.50])

$$iD_\mu \equiv i\partial_\mu - eA_\mu(x), \tag{3.3}$$

where e is the U(1) coupling constant, we write the new Lagrangian as

$$\mathscr{L}(x) = \bar{\psi}(x)(i\not{D} - m)\psi(x). \tag{3.4}$$

It is now easy to see then under the local U(1) transformation

$$\psi(x) \to \exp(i\not{e}\chi(x))\psi(x), \tag{3.5}$$

that, provided that $A_\mu(x)$ becomes

$$A_\mu(x) \to A_\mu(x) - \frac{1}{e}\,\partial_\mu\chi, \tag{3.6}$$

the quantity $iD_\mu\psi(x)$ transforms "covariantly":

$$iD_\mu\psi(x) \to \left(i\partial_\mu - e\left(A_\mu - \frac{1}{e}\,\partial_\mu\chi(x)\right)\right)e^{ie\chi(x)}\psi(x) = e^{ie\chi(x)}(iD_\mu\psi(x)). \tag{3.7}$$

We find then

$$\begin{aligned}\mathscr{L}(x) &= \bar{\psi}(x)(i\gamma^\mu D_\mu - m)\psi(x)\\ &\to \bar{\psi}(x)e^{-ie\chi(x)}e^{ie\chi(x)}(i\gamma^\mu D_\mu - m)\psi(x)\\ &= \bar{\psi}(x)(i\gamma^\mu D_\mu - m)\psi(x) = \mathscr{L}(x),\end{aligned} \tag{3.8}$$

so that this Lagrangian is indeed locally U(1) invariant.

We still must append the free Lagrangian

$$\mathscr{L} = -\frac{1}{4}\,F_{\mu\nu}(x)F^{\mu\nu}(x) \tag{3.9}$$

for the gauge field A_μ. Here

$$F_{\mu\nu}(x) = \partial_\mu A_\nu(x) - \partial_\nu A_\mu(x) \tag{3.10}$$

is the electromagnetic field tensor and is itself invariant under a local U(1) transformation since

$$F_{\mu\nu} \to F_{\mu\nu} - \frac{1}{e}\,\partial_\mu\partial_\nu\chi + \frac{1}{e}\,\partial_\nu\partial_\mu\chi = F_{\mu\nu}. \tag{3.11}$$

(For future reference, it is useful to note that the form of the invariant field tensor can be generated in terms of the commutator of covariant derivatives,

$$[D_\mu, D_\nu] = -ieF_{\mu\nu}). \tag{3.12}$$

The full locally U(1) invariant Lagrangian then becomes

$$\mathscr{L}(x) = \bar{\psi}(x)(i\not{D} - m)\psi(x) - \frac{1}{4}F_{\mu\nu}F^{\mu\nu}, \tag{3.13}$$

which is just the Lagrangian of quantum electrodynamics. (Indeed, by varying the Lagrangian with respect to A_μ we reproduce the Maxwell equations

$$\partial^\mu F_{\mu\nu} = e\bar{\psi}\gamma_\nu\psi). \tag{3.14}$$

Thus, by demanding the *mathematical* requirement of local U(1) invariance, we have produced a Lagrangian that is known to have great *physical* significance. Note also that once we identify the field $A_\mu(x)$ with the electromagnetic vector potential, we learn that its quanta—the photons—*must* be massless, since a mass term

$$\mathscr{L} \sim \frac{1}{2}m_\gamma{}^2 A_\mu A^\mu \tag{3.15}$$

would violate the requirement of local U(1) invariance. Experimentally, this masslessness has been verified to extremely high precision by astrophysical methods, which have yielded an upper bound [2]

$$m_\gamma < 3 \times 10^{-27}\,\text{eV}. \tag{3.16}$$

Buoyed by our success with the use of local U(1) invariance, let's see what happens when we attempt to generalize these techniques to a larger gauge group. Thus, for example, for the special unitary group SU(2), we define the covariant derivative

$$\text{SU(2)}: D_\mu = \partial_\mu - i\frac{g}{2}\boldsymbol{\tau}\cdot\mathbf{W}_\mu, \tag{3.17}$$

while for SU(2) ⊗ U(1) we have

$$\text{SU(2)} \otimes \text{U(1)}: \partial_\mu - i\frac{g}{2}\boldsymbol{\tau}\cdot\mathbf{W}_\mu - i\frac{g'}{2}YB_\mu, \tag{3.18}$$

where g,g' are the appropriate SU(2), U(1) coupling constants, $\mathbf{W}_\mu,B_\mu$ is a triplet, singlet of vector gauge fields, and Y is a hypercharge operator, such that

$$Q = I_3 + \frac{Y}{2}, \tag{3.19}$$

as is the case with the usual isotopic spin and hypercharge.

The commutator

$$[D_\mu,D_\nu] = -ig\frac{1}{2}\boldsymbol{\tau}\cdot\mathbf{F}_{\mu\nu} - ig'\frac{1}{2}YG_{\mu\nu}, \tag{3.20}$$

where

$$\begin{aligned}\mathbf{F}_{\mu\nu} &= \partial_\mu\mathbf{W}_\nu - \partial_\nu\mathbf{W}_\mu + g\mathbf{W}_\mu\times\mathbf{W}_\nu\\ G_{\mu\nu} &= \partial_\mu B_\nu - \partial_\nu B_\mu\end{aligned} \tag{3.21}$$

suggests that one can write a locally SU(2) ⊗ U(1) invariant Lagrangian as

$$\begin{aligned}\mathscr{L} = \bar{D}&\left(i\partial\!\!\!/ + \frac{g}{2}\boldsymbol{\tau}\cdot W\!\!\!\!/ + \frac{g'}{2}Y_D B\!\!\!/\right)D\\ &+\bar{S}\left(i\partial\!\!\!/ + \frac{g'}{2}Y_S B\!\!\!/\right)S - \frac{1}{4}\mathbf{F}_{\mu\nu}\cdot\mathbf{F}^{\mu\nu} - \frac{1}{4}G_{\mu\nu}G^{\mu\nu},\end{aligned} \tag{3.22}$$

where $D(x),S(x)$ is a spinor field which transforms as a doublet, singlet under the SU(2) group. It is then straightforward to see that this form is invariant both under SU(2) rotations,

$$\begin{aligned}D(\mathbf{x}) &\to \exp\left(i\boldsymbol{\alpha}(x)\cdot\boldsymbol{\tau}\frac{1}{2}\right)D(\mathbf{x})\\ S(\mathbf{x}) &\to S(\mathbf{x})\\ \frac{1}{2}\boldsymbol{\tau}\cdot\mathbf{F}_{\mu\nu}(x) &\to \exp\left(i\boldsymbol{\alpha}(x)\cdot\frac{1}{2}\boldsymbol{\tau}\right)\frac{1}{2}\boldsymbol{\tau}\cdot F_{\mu\nu}(x)\exp\left(-i\boldsymbol{\alpha}(x)\cdot\frac{1}{2}\boldsymbol{\tau}\right)\\ &\quad -\frac{i}{g}\left(\partial_\mu\exp\left(i\boldsymbol{\alpha}(x)\cdot\frac{1}{2}\boldsymbol{\tau}\right)\right)\exp\left(-i\boldsymbol{\alpha}(x)\cdot\frac{1}{2}\boldsymbol{\tau}\right)\\ G_{\mu\nu}(x) &\to G_{\mu\nu}(x),\end{aligned} \tag{3.23}$$

and under a hypercharge gauge change

$$D(x) \to \exp\left(i\beta(x)\frac{Y_D}{2}\right)D(x)$$
$$S(x) \to \exp\left(i\beta(x)\frac{Y_S}{2}\right)S(x) \tag{3.24}$$
$$\mathbf{F}_{\mu\nu} \to \mathbf{F}_{\mu\nu} \qquad G_{\mu\nu} \to G_{\mu\nu}.$$

The "art" in this game is choosing the proper representations for the various elementary particles. Focusing first on the leptons, it is conventional to choose $(\nu_l, l)_L$ to form a doublet with respect to the SU(2) group and l_R to be a singlet. (The neutrino, assumed to be massless, has no right-handed component.) The hypercharge eigenvalues can now be determined, since

$$Q = I_3 + \frac{Y}{2} \Rightarrow \begin{matrix} \begin{pmatrix} \nu_l \\ l \end{pmatrix}_L & Y = -1 \\ l_R & Y = -2, \end{matrix} \tag{3.25}$$

with $l = e$, μ, or τ. Summing over the three generations of lepton, we then have as a model Lagrangian

$$\mathscr{L} = \sum_{i=1}^{3} \bar{\psi}_{i_L}\left(i\partial\!\!\!/ + \frac{g}{2}\boldsymbol{\tau}\cdot \mathbf{W}\!\!\!/ - \frac{1}{2}g'B\!\!\!/\right)\psi_{i_L}$$
$$+ \sum_{i=1}^{3} \bar{\psi}_{i_R}(i\partial\!\!\!/ - g'B\!\!\!/)\psi_{i_R} - \frac{1}{4}\mathbf{F}_{\mu\nu}\cdot\mathbf{F}^{\mu\nu} - \frac{1}{4}G_{\mu\nu}G^{\mu\nu}. \tag{3.26}$$

This Lagrangian has some very nice features. Besides its local gauge invariance, we see that the charged W_μ bosons,

$$W_\mu^{\pm} = \frac{1}{\sqrt{2}}(W_\mu^1 \pm iW_\mu^2), \tag{3.27}$$

couple to the lepton currents

$$\sum_i \bar{\psi}_{i_L}\tau_a\gamma_\mu\psi_{i_L} = \begin{pmatrix} \bar{e}_L\gamma_\mu\nu_{e_L} + \bar{\mu}_L\gamma_\mu\nu_{\mu_L} + \bar{\tau}_L\gamma_\mu\nu_{\tau_L} \\ \bar{\nu}_{e_L}\gamma_\mu e_L + \bar{\nu}_{\mu_L}\gamma_\mu\mu_L + \bar{\nu}_{\tau_L}\gamma_\mu\tau_L \end{pmatrix}, \tag{3.28}$$

as required. However, there also exist significant problems with this form. Thus, if we view the weak interaction as taking place via the exchange of

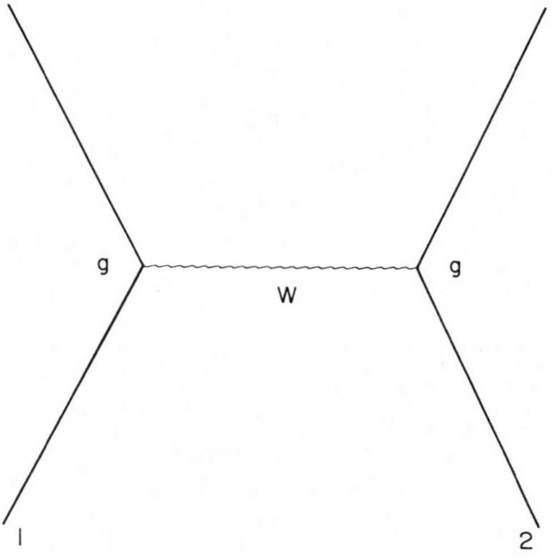

FIGURE 3.1. The charged current component of the weak interaction takes place through the exchange of a W-boson.

the charged W-boson, as shown in Figure 3.1, then the effective four-quark interaction will have the structure

$$\mathscr{L}_W^{\text{eff}} = g^2 J_\mu^{(1)} \frac{1}{q^2 - M_W{}^2} J_{(2)}^\mu. \tag{3.29}$$

In the low energy limit, wherein $q^2 \ll M_W{}^2$, this assumes the usual current-current form

$$\mathscr{H}_W^{\text{eff}} \cong \frac{G}{\sqrt{2}} J_\mu^{(1)} J_{(2)}^\mu, \tag{3.30}$$

where the effective weak coupling G is given by

$$\frac{G}{\sqrt{2}} = \frac{g^2}{M_W{}^2}. \tag{3.31}$$

Using the value $G \cong 10^{-5} m_p^{-2}$ as measured in muon decay $\mu^- \to e^- \bar{\nu}_e \nu_\mu$ and taking $g^2/4\pi \sim \alpha$, where α is the fine structure constant,

$$M_W \sim \sqrt{\frac{4\sqrt{2}\pi\alpha}{G}} \sim 110 \text{ GeV}. \tag{3.32}$$

(This large mass scale for M_W is also suggested by the known short range of the weak interaction. [3]) Thus, the W must be quite massive. However, a canonical mass term of the form

$$\mathscr{L} \sim \frac{1}{2} m_W{}^2 W_\mu W^\mu \tag{3.33}$$

violates SU(2) × U(1) invariance and hence cannot be included in the Lagrangian. In addition, a quark mass term

$$\mathscr{L} \sim m\bar{\psi}\psi = m(\bar{\psi}_L\psi_R + \bar{\psi}_R\psi_L) \tag{3.34}$$

is also not SU(2) × U(1) invariant so we seem to require that $m_e = m_\mu = m_\tau = 0$, in contradiction with experiment.

Obviously, we need a new idea, to give both the gauge bosons and the leptons mass without doing damage to the local gauge invariance. The solution lies in the so-called Higgs mechanism [4]. We introduce a doublet of scalar fields

$$\phi = \begin{pmatrix} \phi^+ \\ \phi^0 \end{pmatrix}: Y = 1 \tag{3.35}$$

and add a gauge-invariant coupling,

$$\mathscr{L} = (D_\mu\phi)^\dagger D^\mu\phi - V(\phi^\dagger\phi), \tag{3.36}$$

to the Lagrangian, where

$$iD_\mu\phi = \left(i\partial_\mu + \frac{g}{2}\boldsymbol{\tau}\cdot\mathbf{W}_\mu + \frac{g'}{2}B_\mu\right)\phi \tag{3.37}$$

and $V(\phi^\dagger\phi)$ is chosen to have a minimum at some nonzero value of its argument. A typical form is (Figure 3.2)

$$V(\phi^\dagger\phi) = -\mu^2\phi^\dagger\phi + \lambda(\phi^\dagger\phi)^2, \tag{3.38}$$

which has its minimum value when

$$\phi^\dagger\phi = \frac{\mu^2}{2\lambda}. \tag{3.39}$$

The symmetry is then assumed broken by the neutral scalar developing a vacuum expectation value

$$\langle 0|\phi|0\rangle = \begin{pmatrix} 0 \\ v/\sqrt{2} \end{pmatrix} \quad \text{where } v = \left(\frac{\mu^2}{\lambda}\right)^{1/2}. \tag{3.40}$$

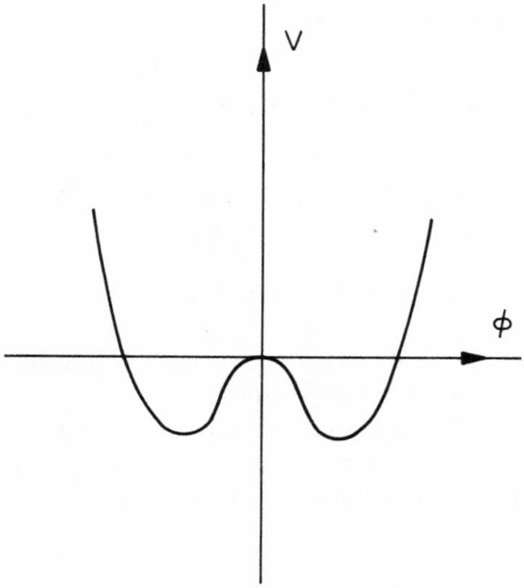

FIGURE 3.2. Plotted is the form of the potential given in eq. (3.38) associated with the phenomenon of spontaneous symmetry breaking. Note that the minimum of the potential energy occurs at a nonzero value of the field strength.

However, the Lagrangian itself is still invariant. The usual example of this phenomenon that is presented is that of ordinary magnetism. No one doubts that the fundamental Hamiltonian describing a piece of iron is rotationally invariant. However, below the Curie temperature a spontaneous magnetization can develop which picks out a particular direction in space. It is not, however, that the fundamental interaction is rotationally noninvariant, but rather that the "vacuum" or ground state does not possess the invariance of the Hamiltonian. This phenomenon is termed "spontaneous symmetry breaking" [4]. In our case we find

$$(D_\mu\phi)^\dagger D_\mu\phi = \frac{1}{2}v^2 \times \left(\frac{1}{4}g'^2 B_\mu{}^2 + \frac{1}{4}g^2\mathbf{W}_\mu{}^2 - \frac{1}{2}gg'B_\mu W^{\mu(0)}\right) + \ldots, \quad (3.41)$$

which can be diagonalized via the definitions

$$\begin{aligned} W_\mu{}^\pm &= \frac{1}{\sqrt{2}}(W_\mu{}^1 \pm iW_\mu{}^2) \\ Z_\mu &= \cos\theta_w W_\mu{}^3 - \sin\theta_w B_\mu \\ A_\mu &= \sin\theta_w W_\mu{}^3 + \cos\theta_w B_\mu, \end{aligned} \quad (3.42)$$

where

$$\tan \theta_w = \frac{g'}{g} \tag{3.43}$$

and θ_w is called the Weinberg angle. In terms of these new fields,

$$(D_\mu\phi)^\dagger D^\mu\phi = \frac{1}{2} M_W{}^2(W_\mu{}^+ W_\mu{}^- + W_\mu{}^- W_\mu{}^+) + \frac{1}{2} M_Z{}^2 Z_\mu Z_\mu + \ldots, \tag{3.44}$$

with

$$M_W{}^2 = \frac{g^2}{4} v^2, \qquad M_Z{}^2 = \frac{g^2 + g'^2}{4} v^2. \tag{3.45}$$

There exists then a relation between the Z and W masses, and the photon field A_μ remains massless

$$M_Z = \frac{M_W}{\cos \theta_w}, \qquad M_A = 0. \tag{3.46}$$

The electroweak coupling becomes

$$\begin{aligned}\mathscr{L} &= \frac{g}{\sqrt{2}}(\bar{\nu}_{e_L}\gamma_\mu e_L W_\mu{}^- + \text{h.c.}) + (e \to \mu) + (\mu \to \tau) \\ &\quad - g \sin \theta_w A_\mu(\bar{e}\gamma_\mu e + \bar{\mu}\gamma_\mu\mu + \bar{\tau}\gamma_\mu\tau) - \frac{g}{\cos \theta_w} J_\mu{}^Z Z_\mu,\end{aligned} \tag{3.47}$$

with

$$J_\mu{}^Z = \sum_i \frac{1}{2} \bar{\psi}_{i_L}\tau_3\gamma_\mu\psi_{i_L} + \sin^2\theta_w(\bar{e}\gamma_\mu e + \bar{\mu}\gamma_\mu\mu + \bar{\tau}\gamma_\mu\tau). \tag{3.48}$$

Thus, we identify the electric charge to be

$$e = g \sin \theta_w \tag{3.49}$$

and the effective weak Hamiltonian as

$$\begin{aligned}H_w^{\text{eff}} &= \frac{g^2}{8M_W{}^2} J_\mu^W J_\mu^{W^\dagger} + \frac{g^2}{\cos^2\theta_w M_Z{}^2} J_\mu{}^Z J_\mu^{Z^\dagger} \\ &\equiv \frac{G}{\sqrt{2}}(J_\mu^W J_\mu^{W^\dagger} + 8J_\mu{}^Z J_\mu^{Z^\dagger}).\end{aligned} \tag{3.50}$$

Here, $G \approx 10^{-5}m_p^{-2}$ is the universal Fermi coupling discussed earlier and the Weinberg angle θ_w has been measured via a variety of neutral current reactions, yielding [5]

$$\sin^2\theta_w \equiv \begin{cases} 0.221 \pm 0.015 & \text{obtained from } e - D \text{ asymmetry} \\ 0.223 \pm 0.003 & \text{obtained from } \nu_\mu N \to \nu_\mu X \\ 0.210 \pm 0.033 & \text{obtained from } \nu_\mu p \to \nu_\mu p \\ 0.223 \pm 0.018 & \text{obtained from } \nu_\mu e \to \nu_\mu e. \end{cases} \tag{3.51}$$

The consistency of these values for $\sin^2\theta_w$ is a strong confirmation of the correctness of the Weinberg-Salam picture. A recent analysis of all available data yields a best-fit value:

$$\sin^2\theta_w = 0.230 \pm 0.005. \tag{3.52}$$

We can now combine the experimental numbers for G, α, θ_w in order to yield specific values for the W and Z masses:

$$\begin{aligned} M_W &= \left(\frac{\pi}{\sqrt{2}G}\frac{\alpha}{\sin^2\theta_w}\right)^{1/2} \approx 82 \text{ GeV} \\ M_Z &= \frac{M_W}{\cos\theta_w} \approx 95 \text{ GeV}. \end{aligned} \tag{3.53}$$

It is well known by now that particles with precisely these masses were detected in 1982 at CERN, winning the Nobel prize for Carlo Rubbia and Simon van der Meer [6].

Thus, the Higgs trick has eliminated one of our problems—generating masses for W_μ and Z_μ. However, what about the need for nonvanishing lepton mass? It turns out that this is solved at the same time, since we can write a gauge invariant coupling of the Higgs doublet to the quark sector,

$$\mathscr{L} = -\sum_i G_i(\bar{\psi}_{i_L}\phi\psi_{i_R} + \bar{\psi}_{i_R}\phi^\dagger\psi_{i_L}), \tag{3.54}$$

which becomes, using the mechanism of spontaneous symmetry breaking,

$$\begin{aligned} \mathscr{L} \sim -\frac{v}{\sqrt{2}}\{&G_1(\bar{e}_L e_R + \bar{e}_R e_L) + G_2(\bar{\mu}_L\mu_R + \mu_R\mu_L) \\ &+ G_3(\bar{\tau}_L\tau_R + \bar{\tau}_R\tau_L)\}, \end{aligned} \tag{3.55}$$

whence we identify

$$m_e = G_1 \frac{v}{\sqrt{2}}, \qquad m_\mu = G_2 \frac{v}{\sqrt{2}}, \qquad m_\tau = G_3 \frac{v}{\sqrt{2}} \tag{3.56}$$

and

$$m_{\nu_e} = m_{\nu_\mu} = m_{\nu_\tau} = 0. \tag{3.57}$$

Now that the leptons are in satisfactory shape, we consider how the formalism can be modified to handle the generation of quark masses. In order to do this, we recall that since there exists only a single two-dimensional representation of SU(2)—i.e., the iso-doublet—then if ϕ is a doublet, so must its conjugate be. That is,

$$\begin{aligned} \phi &= \begin{pmatrix} \phi^+ \\ \phi^0 \end{pmatrix} : Y = 1 \\ \tilde{\phi} &= i\sigma_2 \phi = \begin{pmatrix} \phi^0 \\ -\phi^- \end{pmatrix} : Y = -1 \end{aligned} \tag{3.58}$$

are both iso-doublets.

For simplicity, assume initially that there exist only two quark generations,

$$\begin{aligned} &\begin{pmatrix} u \\ d \end{pmatrix}_L \begin{pmatrix} c \\ s \end{pmatrix}_L : Y = \frac{1}{3} \\ &u_R, d_R \; c_R, s_R : Y = \frac{4}{3}, -\frac{2}{3}. \end{aligned} \tag{3.59}$$

Then we can write an $SU(2)_L \otimes U(1)$ invariant Lagrangian as

$$\begin{aligned} \mathscr{L} &= \sum_i \bar{\psi}_{i_L} \left(i\not{\partial} + \frac{g}{2} \tau \cdot \not{W} + \frac{g'}{6} \not{B} \right) \psi_{i_L} \\ &\quad + \sum_i \left(\bar{\psi}^u_{iR} \left(i\not{\partial} + \frac{2}{3} g' \not{B} \right) \psi^u_{iR} + \bar{\psi}^d_{iR} \left(i\not{\partial} - \frac{1}{3} g' \not{B} \right) \psi^d_{iR} \right), \end{aligned} \tag{3.60}$$

in an obvious notation. We see that the $W^\pm$ now couple to the current

$$\bar{u}_L \gamma_\mu d_L + \bar{c}_L \gamma_\mu s_L, \tag{3.61}$$

which looks promising, except that the Cabibbo angle is missing: the piece of the current involving the u quark should have the form

$$\bar{u}_L\gamma_\mu(d_L \cos\theta_c + s_L \sin\theta_c). \tag{3.62}$$

To see how this quark mixing arises, recall that we have yet to generate quark masses. This can be accomplished using the Higgs trick by writing down general $SU(2)_L \times U(1)$ invariant Yukawa couplings of the type

$$\mathscr{L} = \sum_{i,j} \Gamma^d_{ij}\bar{\psi}_{iL}\phi\psi^d_{jR} + \Gamma^u_{ij}\bar{\psi}_{iL}\tilde{\phi}\psi^u_{jR} + \text{h.c.} \tag{3.63}$$

When spontaneous symmetry breaking takes place, we have

$$\phi = \begin{pmatrix} 0 \\ v/\sqrt{2} \end{pmatrix} + \ldots, \qquad \tilde{\phi} = \begin{pmatrix} v/\sqrt{2} \\ 0 \end{pmatrix} + \ldots \tag{3.64}$$

and

$$\mathscr{L} = \sum_{i,j} \frac{v}{\sqrt{2}} \Gamma^d_{ij}\bar{\psi}^d_{iL}\psi^d_{jR} + \frac{v}{\sqrt{2}} \Gamma^u_{ij}\bar{\psi}^u_{iL}\psi^u_{jR} + \text{h.c.}, \tag{3.65}$$

which has the correct form for a mass term, except that it is not diagonal in quark flavors. The diagonalization can easily be achieved by separate unitary transformations within the up and down quark sectors

$$\begin{aligned} \begin{pmatrix} u'_1 \\ u'_2 \end{pmatrix}_{L,R} &= U^u_{L,R} \begin{pmatrix} u_1 \\ u_2 \end{pmatrix}_{L,R} \\ \begin{pmatrix} d'_1 \\ d'_2 \end{pmatrix}_{L,R} &= U^d_{L,R} \begin{pmatrix} d_1 \\ d_2 \end{pmatrix}_{L,R} \end{aligned} \tag{3.66}$$

The mass term then becomes

$$\mathscr{L} = \sum_{i,j} \frac{v}{\sqrt{2}} \tilde{\Gamma}^d_{ij}\bar{\psi}^{d'}_{iL}\psi^{d'}_{jR} + \frac{v}{\sqrt{2}} \tilde{\Gamma}^u_{ij}\bar{\psi}^{u'}_{iL}\psi^{u'}_{jR} + \text{h.c.}, \tag{3.67}$$

with

$$\begin{aligned} \tilde{\Gamma}^d &= U_L{}^d\Gamma^d U_R^{d-1} \\ \tilde{\Gamma}^u &= U_L{}^u\Gamma^u U_R^{u-1} \end{aligned} \tag{3.68}$$

being diagonal mass matrices for d-type, u-type quarks, respectively. The primed states then are the ones we identify with the usual quark eigenstates, and in terms of these primed states, the charged weak current becomes

$$J_\mu{}^W = (\bar{u}\bar{c})_L \gamma_\mu \begin{pmatrix} d \\ s \end{pmatrix}_L = (\bar{u}'\bar{c}')_L \gamma_\mu U \begin{pmatrix} d' \\ s' \end{pmatrix}_L \tag{3.69}$$

where

$$U = U_L{}^u U_L^{d-1} \tag{3.70}$$

is a 2 ⊗ 2 unitary matrix, whose most general form is

$$U = \begin{pmatrix} \cos\theta & \sin\theta \\ -\sin\theta & \cos\theta \end{pmatrix}. \tag{3.71}$$

Hence, we identify θ with the Cabibbo angle θ_c, and observe how naturally mixing occurs in this model. In the so-called standard model that we have been developing, the angle θ_c is merely an arbitrary parameter. However, many theorists have imposed additional constraints on the theory in order to attempt to predict θ_c in terms of quark masses, etc. [7] None of these so-called horizontal symmetry proposals is yet completely convincing.

Although it is not directly relevant to nuclear physics, let me point out that in reality, of course, there exist (at least) three quark generations (although the t quark has not yet been detected),

$$\begin{pmatrix} u \\ d \end{pmatrix} \begin{pmatrix} c \\ s \end{pmatrix} \begin{pmatrix} t \\ b \end{pmatrix}, \tag{3.72}$$

and the weak charged current is written in terms of a general 3 ⊗ 3 unitary matrix, which is usually parametrized in the form given by Kobayashi and Maskawa [8]

$$U = \begin{pmatrix} c_1 & s_1 c_3 & s_1 s_3 \\ -s_1 c_2 & c_1 c_2 c_3 - s_2 s_3 e^{i\delta} & c_1 c_2 s_3 + s_2 c_3 e^{i\delta} \\ -s_1 s_2 & c_1 s_2 c_3 + c_2 s_3 e^{i\delta} & c_1 s_2 s_3 - c_2 c_3 e^{i\delta} \end{pmatrix}, \tag{3.73}$$

where here $c_i, s_i = \cos\theta_i, \sin\theta_i$ and δ is an arbitrary phase, which, if different from 0 or π, will lead to CP noninvariance. Much effort has gone into the determination of the three KM angles, and as of now it appears that the KM picture may be able to account not only for the charged weak interactions of quarks, but also for the mysterious CP violation observed in the $K_L - K_S$ system [9], as we shall discuss in a later chapter.

The Weinberg-Salam model described above has thus proved incredibly successful in explaining an enormous variety of weak and electromagnetic phenomena and may be said to be strongly confirmed. However, one important conundrum remains—where is the so-called Higgs boson? Earlier we characterized the scalar field $\phi(x)$ in terms of its vacuum expectation value v, and this was sufficient to generate masses for the gauge bosons and fermions. However, there are also quantum fluctuations about this classical value. Writing these as

$$\phi(x) = \begin{pmatrix} 0 \\ v/\sqrt{2} \end{pmatrix} + \begin{pmatrix} \phi^+(x) \\ \phi^0(x) \end{pmatrix}, \tag{3.74}$$

we might expect there to exist four such degrees of freedom, associated with the two complex scalar fields $\phi^+(x)$ and $\phi^0(x)$. However, this is illusory, as we can see by parameterizing the field doublet instead via

$$\phi(x) = \exp\left(i\frac{1}{v}\boldsymbol{\tau}\cdot\boldsymbol{\chi}(x)\right)\sqrt{\frac{1}{2}}\begin{pmatrix} 0 \\ v + h(x) \end{pmatrix}. \tag{3.75}$$

There are still four scalar degrees of freedom, characterized by the hermitian fields $\boldsymbol{\chi}(x), h(x)$. However, the triplet $\boldsymbol{\chi}(x)$ can be eliminated by means of the SU(2) gauge transformation

$$\phi(x) \to \phi'(x) = \exp\left(-i\frac{1}{v}\boldsymbol{\tau}\cdot\boldsymbol{\chi}(x)\right)\phi(x). \tag{3.76}$$

The Lagrangian is now independent of $\boldsymbol{\chi}(x)$ (but *not* of $h(x)$) and only a single scalar field remains. This scalar particle is called the Higgs boson and it is clear from eq. (3.75) that the Higgs particle couples to the fundamental Fermi (and gauge Bose) fields in the same way as does the vacuum expectation value v—i.e., proportional to the mass of the appropriate field. The mass of the Higgs itself is undetermined, since it depends on the detailed structure of the potential function $V(\phi^\dagger\phi)$, although various theoretical arguments suggest that it should be in the range [10]

$$10 \text{ GeV} \lesssim m_h \lesssim 300 \text{ GeV}.$$

Despite years of search, no evidence for the existence of the Higgs has yet turned up. This could well be associated with the feature of its coupling to quarks via their mass, and the fact that available targets are rich in u,d but poor in heavier quark species. In any case, the Higgs particle is probably the least satisfactory aspect of the electroweak theory.

We finish this section by clearing up the mystery of what happened to the three scalar degrees of freedom associated with the fields $\chi(x)$ that have been gauged away. The answer may be found by simple counting. In the SU(2) × U(1) model before spontaneous symmetry breaking there existed

(1) Four massless gauge bosons with two degrees of freedom each
(2) Four scalar fields with a single degree of freedom each

for a total of twelve degrees of freedom. On the other hand, after spontaneously symmetry breaking, we find

(1) Three massive gauge bosons with three degrees of freedom each
(2) One massless gauge boson with two degrees of freedom
(3) One massive scalar boson with a single degree of freedom.

The total is still twelve as it must be, and it is now clear what happened to the three extra scalar fields in the unbroken version of the model. The gauge fields have "eaten" the scalars in order to become massive. These scalar degrees of freedom have become the longitudinal polarization states of the three massive gauge bosons.

There is much more we could discuss at this point. However, it's time to discard theoretical manipulations and move to aspects of the electroweak model as manifested in experiments with nuclei.

References

[1] S. Weinberg, Phys. Rev. Lett. *19*, 1264 (1967), and *27*, 1688 (1971); A. Salam in "Elementary Particle Theory: Relativistic Groups and Analyticity (Nobel Symposium No. 8)," ed. by N. Svartholm, Almqvist, and Wilsell, Stockholm (1968), p. 367.
[2] Particle Data Group, Phys. Lett. *170B* (1986).
[3] E. D. Commins, *Weak Interactions*, McGraw-Hill, New York (1973), Ch. 1.
[4] P. W. Higgs, Phys. Lett. *12*, 132 (1964), Phys. Rev. Lett. *13*, 508 (1964), and Phys. Rev. *145*, 1156 (1966).
[5] An excellent summary of all available data is given by U. Amaldi et al., Phys. Rev. *D36*, 1385 (1987).

[6] G. Arnison et al., Phys. Lett. *122B*, 103 (1983), and Phys. Lett. *126B*, 398 (1983); M. Banner et al., Phys. Lett. *122B*, 476 (1983); P. Bagnaia et al., Phys. Lett. *129B*, 130 (1983).
[7] See, e.g., S. Pakvasa and H. Sugawara, Phys. Lett. *82B*, 105 (1979); Y. Yamanaka, H. Sugawara, and S. Pakvasa, Phys. Rev. *D25*, 1895 (1982).
[8] M. Kobayashi and T. Maskawa, Prog. Theor. Phys. *49*, 652 (1973).
[9] C. R. Christenson et al., Phys. Rev. Lett. *13*, 138 (1964).
[10] The upper bound comes from the requirement that perturbation theory be applicable. The lower bound comes from the requirement that one-loop corrections not cause the effective potential to become unstable. See A. D. Linde, JETP Lett. *23*, 64 (1976); S. Weinberg, Phys. Rev. Lett. *36*, 294 (1976).

Chapter 4

SYMMETRIES OF THE CHARGED WEAK CURRENT

Tiger! Tiger! burning bright
In the forests of the night,
What immortal hand or eye
Could frame thy fearful symmetry?
—WILLIAM BLAKE

4.1 Conserved Vector Current

The left-handed nature of the charged weak current $J_\mu{}^W$ and its relation to the electromagnetic current was postulated in 1958 by Feynman and Gell-Mann [1] long before the advent of the Weinberg-Salam model. Their proposal was based partly on experimental and partly on esthetic grounds, and it is interesting that Feynman, in his book *Surely You're Joking, Mr. Feynman,* considers this one of his premier achievements! [2] Nevertheless, as we shall see, this "conserved vector current hypothesis" is inescapable in a quark picture of the weak interaction.

The point is, as discussed previously in the case of the Higgs doublet, both u,d and the corresponding antiquarks $\bar{u},\bar{d}$ must transform as isospinors under SU(2) since there exists a unique two-dimensional irreducible representation. Usually these doublets are chosen to be

$$\begin{pmatrix} u \\ d \end{pmatrix} \quad \text{and} \quad \begin{pmatrix} \bar{d} \\ -\bar{u} \end{pmatrix}. \tag{4.1}$$

Then, commuting the electromagnetic current J_μ^{em} with the isotopic spin-lowering operator $I_- = \int d^3x \bar{d}\gamma_0 u$,

$$\begin{aligned} [I_-, J_\mu^{em}] &= \left[I_-, \frac{2}{3}\,\bar{u}\gamma_\mu u - \frac{1}{3}\,\bar{d}\gamma_\mu d + \dots \right] \\ &= \frac{2}{3}\,\bar{d}\gamma_\mu u + \frac{1}{3}\,\bar{d}\gamma_\mu u = \bar{d}\gamma_\mu u. \end{aligned} \tag{4.2}$$

We observed a quantity of precisely this form in the previous chapter—the *polar* vector component of the charged weak current. Thus, we have derived the so-called conserved vector current hypothesis,

$$[I_\pm, J_\mu^{em}] = \mp V_\mu^{W\pm}, \tag{4.3}$$

which relates matrix elements of the charged weak polar vector current with corresponding electromagnetic amplitudes.

This relation can be, and has been, well tested by experiments involving nuclear beta decay. To see how this has been done, however, it is necessary to introduce some notation. Consider first a transition between $J^p = \frac{1}{2}^+$ systems, e.g., nucleon beta decay. Then from Lorentz invariance alone, the matrix element must have the form [3]

$$\begin{aligned}\langle\beta|J_\mu{}^W|\alpha\rangle = \bar{u}(p_2)\bigg[\gamma_\mu(g_V + g_A\gamma_5) \\ - \frac{i}{m_1 + m_2}\sigma_{\mu\nu}q^\nu(g_M + g_T\gamma_5) \\ + \frac{q_\mu}{m_1 + m_2}(g_S - g_P\gamma_5)\bigg]u(p_1),\end{aligned} \tag{4.4}$$

where

$$q = p_1 - p_2 \tag{4.5}$$

is the momentum transfer and $\bar{u}(p_2)$ and $u(p_1)$ are free Dirac spinors. The structure constants or "form factors" $g_V, g_A, \ldots$ are functions of q^2 and are generally denoted by names that characterize their interaction:

g_V	Fermi or Vector
g_A	Gamow-Teller or Axial Vector
g_M	Weak Magnetism
g_T	Induced Tensor or Weak Electricity
g_S	Induced Scalar
g_P	Induced Pseudoscalar

Here the term "induced" refers to the feature that such couplings are *not* present in the fundamental weak current, which is of the simple

$$\bar{q}\gamma_\mu(1 + \gamma_5)q \tag{4.6}$$

form, but rather appear when wavefunctions between states of confined quarks are taken. For example, in the case of neutron beta decay, the leading vector, axial couplings have the familiar values

$$g_V(q^2 = 0) = 1 \qquad g_A(q^2 = 0) = 1.25. \tag{4.7}$$

However, for our purposes we require matrix elements of the weak current for more general spin sequences appropriate to nuclear beta decay.

For the case of an arbitrary "allowed" ($\Delta J = 0, \pm 1$, no parity change) nuclear beta transition, the definitions of form factors given in eq. (4.4) can be generalized to become [4]

$$l^\mu \langle \beta | V_\mu | \alpha \rangle = \delta_{JJ'} \delta_{MM'} \left(a(q^2) \frac{P \cdot l}{2M} + e(q^2) \frac{q \cdot l}{2M} \right) + ib(q^2) \frac{1}{2M} C^{M'k;M}_{J'1;J} (\mathbf{q} \times \mathbf{l})_k$$
$$+ C^{M'k;M}_{J'2;J} \left[\frac{1}{2M} f(q^2) C^{nn';k}_{11;2} l_n q_{n'} + \frac{1}{(2M)^3} g(q^2) P \cdot l \sqrt{\frac{4\pi}{5}}\, Y_2{}^k(\mathbf{q}) \right] \tag{4.8a}$$

$$l^\mu \langle \beta | A_\mu | \alpha \rangle = C^{M'k;M}_{J'1;J} \varepsilon_{ijk} \varepsilon_{ij\lambda\eta} \frac{1}{4M} \left[c(q^2) l^\lambda P^\eta - d(q^2) l^\lambda q^\eta \right.$$
$$\left. + \frac{1}{(2M)^2} h(q^2) q^\lambda P^\eta q \cdot l \right]$$
$$+ C^{M'k;M}_{J'2;J} C^{nn';k}_{12;2} l_n \sqrt{\frac{4\pi}{5}}\, Y^{n'}_2(\mathbf{q}) \frac{1}{(2M)^2} j_2(q^2)$$
$$+ C^{M'k;M}_{J'3;J} C^{nn';k}_{12;3} l_n \sqrt{\frac{4\pi}{5}}\, Y^{n'}_2(\mathbf{q}) \frac{1}{(2M)^2} j_3(q^2), \tag{4.8b}$$

where we have here assumed a transition between states with spin J, J' and projections M, M' along some axis of quantization, and have used the definitions

$$P_\mu = (p_1 + p_2)_\mu \quad \text{and} \quad M = \frac{1}{2}(M_1 + M_2). \tag{4.9}$$

In eq. (4.8) the meaning of the form factors $a, b, \ldots h$ is analogous to that in eq. (4.4), provided we make the identifications

$$\begin{aligned} a &= g_V & c &= \sqrt{3} g_A \\ b - a &= \sqrt{3} g_M & d &= \sqrt{3} g_T \\ e &= g_S & h &= \sqrt{3} g_P. \end{aligned} \tag{4.10}$$

Of course, the form factors f, g, j_2, j_3, which are associated with tensor operators of rank 2,3, have no $\frac{1}{2}^+ - \frac{1}{2}^+$ analog.

We shall return later to the axial piece of the current, eq. (4.8b). For the moment we focus on the polar vector component, eq. (4.8a), and ask what the CVC relation, eq. (4.3), predicts for the vector structure constants. In the case of a transition between nuclear isotopic analog states,

$$\langle I,I_z \pm 1|V_\mu{}^W|I,I_z\rangle, \tag{4.11}$$

we find, by taking matrix elements of eq. (4.3),

$$\begin{aligned} a(0) &= [(I \mp I_z)(I \pm I_z + 1)]^{1/2} \\ b(0) &= a(0)\sqrt{\frac{J+1}{J}}\,(\mu_\beta - \mu_\alpha) \\ e(0) &= f(0) = 0 \\ g(0) &= -a(0)\left(\frac{(J+1)(2J+3)}{J(2J-1)}\right)^{1/2}\frac{2M^2}{3}\,(Q_\beta - Q_\alpha), \end{aligned} \tag{4.12}$$

where μ,Q are the nuclear magnetic, quadrupole moments, respectively. Thus in the case of neutron beta decay—$I = \frac{1}{2}$, $I_z = -\frac{1}{2}$—we find the familiar results

$$\begin{aligned} a(0) &= 1 \\ \frac{1}{\sqrt{3}}\,b(0) &= \mu_p - \mu_n = 4.7 \\ e(0) &= 0. \end{aligned} \tag{4.13}$$

A classic test of these predictions employs $0^+ - 0^+$ analog beta decay, wherein the spin-parity sequence permits *only* the matrix elements a,e to contribute. Generally, such decay rates are parameterized in terms of the so-called ft value,

$$ft \equiv t_{1/2} \times f \tag{4.14}$$

where $t_{1/2}$ is the half life, and

$$f = \int_{m_e}^{E_{\max}} dE\, F(\pm Z,E)pE(E_0 - E)^2 \tag{4.15}$$

accounts for the phase space. Here $F(\pm Z,E)$ is called the "Fermi function" and corrects the phase space for the feature that the outgoing electron/positron interacts with the remaining nucleus electromagnetically, and therefore is not a plane wave state. (For a point nucleus the Fermi function is simply given by the absolute square of the Coulomb wave function at the origin

$$F(\pm Z,E) = |\psi(0)|^2 = \frac{2\pi\gamma}{e^{2\pi\gamma} - 1} \quad \text{with } \gamma = \frac{\pm Z\alpha E}{p}. \tag{4.16}$$

However, the finite size of an actual nucleus makes the form of a realistic Fermi function much more complex [5].) We then obtain

$$ft = \pi^3 ln\, 2 \frac{1}{G^2 \cos^2\theta_c}, \tag{4.17}$$

which should be identical for *all* such transitions. Actually, our discussion is somewhat cavalier and has neglected finite nuclear size, radiative corrections, etc. Wilkinson has suggested use of a modified phase space factor [6],

$$f^R = f \times (1 + \text{finite size} + \text{radiative corrections} + \ldots), \tag{4.18}$$

in order to account for these small but important terms. Recently, a long-standing problem in the correction at $O(Z\alpha^2)$ was resolved, resulting in a very consistent set of $f^R t$ values for analog $0^+ - 0^+$ decays all the way from $^{14}O(Z = 8)$ to ^{54}Co$(Z = 27)$ as shown below [7]:

	E_0 (KeV)	$f^R t$ (sec)	
^{14}O	1809	3074.0 ± 3.9	
^{26}Al	3211	3068.1 ± 3.7	$f^R t = 3070.6 \pm 1.6$
^{34}Cl	4470	3069.0 ± 4.7	
^{38}K	5023	3066.6 ± 4.6	(4.19)
^{42}Sc	5402	3077.5 ± 7.5	$\chi^2/\nu = 0.57$
^{46}V	6029	3074.7 ± 7.5	
^{50}Mn	6610	3069.6 ± 5.7	
^{54}Co	7220	3070.6 ± 1.6	

Obviously the fit is excellent, and to this level of accuracy—0.2%—we may say that the data strongly confirm the CVC prediction. Actually, we can go even further and check the overall normalization, since from analysis of muon decay it is possible to determine the Fermi coupling,

$$\mu \text{ decay: } \frac{G^{(0)}}{\sqrt{2}}, \tag{4.20}$$

while from nuclear beta decay one can evaluate the Fermi coupling multiplied by the cosine of the Cabibbo angle,

$$\beta\text{-decay: } \frac{G^{(0)}}{\sqrt{2}} \cos\theta_c. \tag{4.21}$$

At this level of precision, however, it is necessary to include also electromagnetic renormalization of these weak couplings, yielding

$$G_\mu^{\exp} = G_\mu^{(0)}(1 + \frac{1}{2}\Delta_\mu{}^\alpha)$$
$$G_\rho^{\exp} = G_\beta^{(0)}(1 + \frac{1}{2}\Delta_\beta{}^\alpha). \tag{4.22}$$

The radiative corrections $\Delta_\mu{}^\alpha$, $\Delta_\beta{}^\alpha$ have been calculated by Sirlin as [8]

$$\Delta_\beta{}^\alpha - \Delta_\mu{}^\alpha = \frac{\alpha}{2\pi}\left\{4\, ln\frac{M_Z}{m_N} + ln\frac{m_N}{m_l} + 2C + A_g\right\}. \tag{4.23}$$

Here A_g represents a small but calculable $O(\alpha_s)$ perturbative QCD correction, while $ln\frac{m_N}{m_A} + 2C$ is an axial-vector induced, structure-dependent contribution, m_A being a low-energy cutoff applied to the short distance part of the $\gamma - W$ box diagram [9]. This cutoff is expected to be in the 1 GeV range and $2C$ is the remaining long-distance component. A recent estimate gives $C \cong 0.9$ [10]. Using this radiative correction factor, eq. (4.19) for the ft value of analog $0^+ - 0^+$ decays, and the experimental muon decay rate

$$\Gamma_\mu = (2.99554 \pm 0.00005) \times 10^{-16}\ \text{MeV}, \tag{4.24}$$

we find

$$\cos\theta_1 = 0.9744 \pm 0.0010. \tag{4.25}$$

Taking [11]

$$\sin\theta_1 \cos\theta_3 = 0.221 \pm 0.002 \tag{4.26}$$

from analysis of $K \to \mu\nu$ and strangeness changing hyperon decay, we find the upper bound

$$|s_3| < 0.2. \tag{4.27}$$

It is remarkable that we can place such a strong bound on an angle measuring *heavy* quark couplings from experiments at the lowest energies.

It is important to note that this *ft* analysis actually is probing matrix elements of the weak charge (i.e., the *EO* moment of the current),

$$Q^W = \int d^3r J_0{}^W(\mathbf{r},t). \tag{4.28}$$

It is also possible to test the CVC condition by measurement of the M1 moment of the current as first suggested by Gell-Mann [12]. The basic idea is to measure the energy spectrum of the emitted electron or positron and to compare with the result expected from the conserved vector current hypothesis. Defining the shape factor $S(E)$ via

$$\frac{d\Gamma}{dE} \propto \text{Phase Space} \times S(E), \tag{4.29}$$

where

$$\text{Phase Space} \sim F(\pm Z,E)pE(E_0 - E)^2 \tag{4.30}$$

is a kinematic factor, then in simplest approximation

$$S(E) = 1 \pm \frac{4E}{3M} b/c \qquad \text{for } \begin{matrix} e^- \\ e^+ \end{matrix} \text{ decay} \tag{4.31}$$

in a "Gamow-Teller" (i.e., nonanalog) transition. Ideally, one uses a pair of mirror nuclei and can measure both the β^+ and β^- shapes, which can eliminate certain systematic uncertainties. Gell-Mann suggested using the $A = 12$ isotriplet—^{12}B, ^{12}C(15.11 MeV, 1^+), ^{12}N—(cf. Figure 4.1) wherein the weak magnetism structure function b can be predicted in terms of the radiative M1 width of the analog ^{12}C state using the conserved vector current assumption,

$$b = \left(\frac{\Gamma_{M1} \cdot 6M^2}{\alpha E_0{}^3}\right)^{1/2}, \tag{4.32}$$

where E_0 is the Q-value for the radiative transition. Of course, in reality, things are not that simple. First, the effect is quite small; assuming

$$\left(\frac{b}{Ac}\right)_{\text{nuclear}} \sim \left(\frac{b}{Ac}\right)_{\text{neutron}} \sim 4.7, \tag{4.33}$$

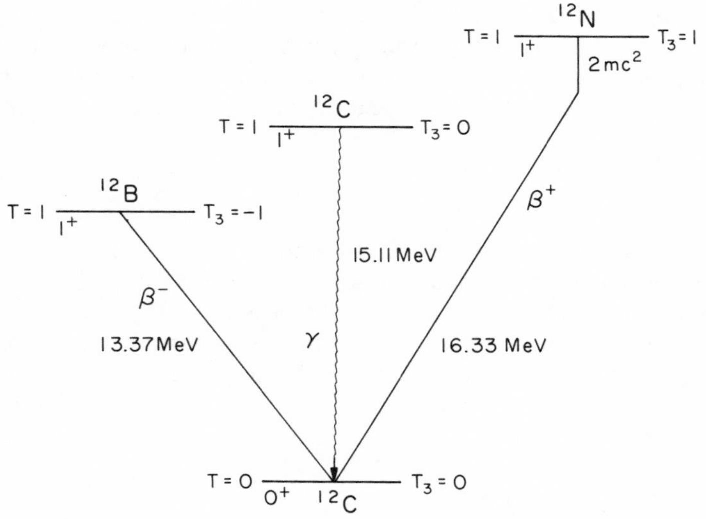

FIGURE 4.1. Indicated is the famous isotopic triplet of 1^+ states in the $A = 12$ system whose β,γ decays to the ^{12}C ground state are used as a test of the conserved vector current hypothesis.

we find

$$\frac{dS}{dE} \simeq \pm\frac{4}{3}\frac{b}{Ac}\frac{1}{m_N} \sim \pm 0.5\%/\text{MeV}, \tag{4.34}$$

a *very* small slope requiring considerable experimental skill to measure. Second, a realistic theoretical analysis of the shape factor must include radiative corrections, finite size effects, etc., yielding [13]

$$\begin{aligned} S(E) = \Bigg\{ & 1 + \frac{2}{9}\frac{c'(0)}{c(0)}(11m_e^2 + 20EE_0 - 20E^2 - 2m_e^2E_0/E) \\ & - \frac{c'(0)}{c(0)}\left(\frac{9}{2}\left(\frac{\alpha Z}{R}\right)^2 \mp \frac{2}{3}\frac{\alpha Z}{R}E_0 \pm \frac{20}{3}\frac{\alpha Z}{R}E\right) \\ & + \frac{\sqrt{10}}{6}\frac{\alpha Z}{MR}\left(2\frac{b}{c} \pm \frac{d}{c} \pm 1\right) - \frac{2}{3}\frac{E_0}{M}\left(1 + \frac{d}{c} \pm \frac{b}{c}\right) \\ & + \frac{2}{3}\frac{E}{M}\left(5 \pm 2\frac{b}{c}\right) - \frac{m_e^2}{3ME}\left(2 + \frac{d}{c} \pm 2\frac{b}{c} - \frac{h}{c}\frac{E_0 - E}{M}\right)\Bigg\} \end{aligned} \tag{4.35}$$

in a more careful discussion.

Nevertheless, experimental measurements have been undertaken by a number of groups. One of the earliest of these was the measurement in the $A = 12$ system by Lee, Mo, and Wu using an iron-free magnetic spectrometer [14]. (Although certain problems were pointed out with the analysis of the original experiment [15], a corrected study was performed, yielding the results quoted below [16].) More recently, the same $A = 12$ system was examined by Kaina et al. using a NaI scintillation spectrometer [17]. The results of these experiments are summarized as follows:

			Exp	*%/MeV Theory*
$\frac{dS}{dE}$	Lee et al. [16]	$^{12}B \to {}^{12}C$	0.48 ± 0.10	0.43
	Kaina et al. [17]		0.91 ± 0.11	
	Lee et al.	$^{12}N \to {}^{12}C$	-0.52 ± 0.06	-0.50
	Kaina et al.		-0.07 ± 0.09	

(4.36)

The Lee, Mo, Wu measurement is in good agreement with the CVC prediction for both branches, as is clear both from the average values of the slopes given in eq. (4.35) and from the detailed spectral shape shown in Figure 4.2. In the case of the Kaina et al. experiment, both positron and

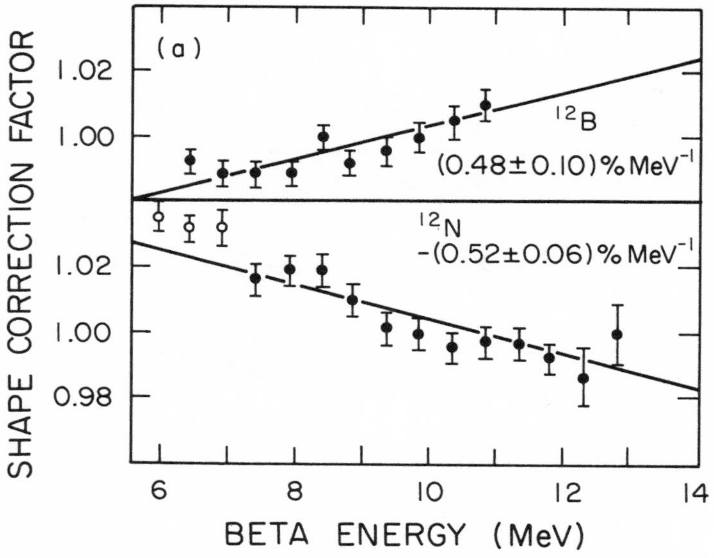

FIGURE 4.2. Shown are the "narrow slit" shape factors measured with an iron-free spectrometer by Lee, Mo, and Wu, ref. 14, as reanalyzed by Wu, Lee, and Mo, ref. 16.

electron slopes appear to disagree with the theoretical expectation. However, if we take the ratio, in which various systematic effects are eliminated, we get good agreement:

	Expt.	*%/MeV*	*Theory*	
$\frac{dS^-}{dE} - \frac{dS^+}{dE} =$	1.00 ± 0.13	Lee et al.	0.93	(4.37)
	0.98 ± 0.14	Kaina et al.		

An Argonne/Los Alamos group has also done some preliminary work in the $A = 12$ system, but results are not yet available.

Similar shape factor experiments were performed in the $A = 20$ system by Calaprice and Alburger [18], by Genz et al. [19], and by van Elmbt et al. [20]. However, only the electron branch $^{20}F \rightarrow {}^{20}Ne$ was studied and the results, although somewhat less precise than those in $A = 12$, are in general agreement with the CVC expectation:

		Expt.	*%/MeV*	*Theory*	
$\frac{dS}{dE} =$	Genz et al.	$^{20}F \rightarrow {}^{20}Ne$	0.78 ± 0.43	0.94	(4.38)
	Calaprice and Alburger		0.4 ± 0.5		
	van Elmbt et al.		1.12 ± 0.07	1.18	

Finally, we note that it is also possible to use the data from $0^+ - 0^+$ analog decays in order to place limits on the size of a possible induced scalar coupling $e(q^2)$, which is required to vanish according to CVC eq. (4.13). The point is that, if such a term were present, it would show up as a systematic deviation of the ft values from constancy

$$ft = 2\pi^3 \ln 2 \frac{1}{G^2 \cos^2\theta_c} \frac{1}{a^2 + ae \left\langle \frac{m_e^2}{ME} \right\rangle}. \tag{4.39}$$

The spectral average value $\left\langle \frac{1}{E} \right\rangle$ depends on the Q-value for the decay in question, which is itself dependent on the Coulomb splitting between such levels, and hence on Z. There is no evidence for such an effect in the data, and from a fit to these ft values one obtains the limit [21]

$$\frac{1}{A} e = -3 \pm 8. \tag{4.40}$$

This bound is somewhat crude because of the feature that the induced scalar always appears in spectra accompanied by the kinematic factor $m_e{}^2/m_N E \ll 1$. Some advantage may be gained by examination of muon capture data, when the kinematic factors become $m_\mu/m_N \sim 0.1$. However, this is accomplished only at the cost of considerable model-dependent assumptions and will not be discussed here [22].

We conclude then that an entire series of nuclear beta decay experiments are in basic support of the CVC hypothesis, but additional and more precise data would be welcome.

4.2 Second-Class Currents

Another symmetry of the weak current that has been probed via nuclear beta decay is that of its behavior under the product of charge conjugation, which interchanges particles and antiparticles, coupled with a 180° rotation about the Y-axis in isotopic spin space. This combination is conventionally called "G-parity" and is a symmetry of the strong interactions:

$$G \equiv C \exp(-i\pi I_2). \tag{4.41}$$

Now, under charge conjugation, we have

$$C\bar{u}\gamma_\mu(1 + \gamma_5)dC^{-1} = -\bar{d}\gamma_\mu(1 - \gamma_5)u. \tag{4.42}$$

However, under an isotopic spin rotation,

$$\exp(-i\pi I_2)\begin{pmatrix} u \\ d \end{pmatrix} = -i\sigma_2\begin{pmatrix} u \\ d \end{pmatrix} = \begin{pmatrix} -d \\ u \end{pmatrix}. \tag{4.43}$$

Thus we find

$$G\bar{u}\gamma_\mu(1 + \gamma_5)dG^{-1} = \bar{u}\gamma_\mu(1 - \gamma_5)d, \tag{4.44}$$

and we observe that it is an inescapable conclusion of the quark model that the charged weak $\Delta S = 0$ current transforms under G as

$$\begin{aligned} GV_\mu G^{-1} &= V_\mu \\ GA_\mu G^{-1} &= -A_\mu. \end{aligned} \tag{4.45}$$

In principle, one could imagine currents with the *opposite* G-symmetry properties,

$$\begin{aligned} GV_\mu G^{-1} &= -V_\mu \\ GA_\mu G^{-1} &= A_\mu, \end{aligned} \tag{4.46}$$

although such objects lie outside this *conventional* quark picture. For example, an explicit (toy) model for such currents could be constructed by assuming the existence of *two* pairs of *u,d* quarks which differ by some unobserved quantum number and which both transform as doublets under a strong SU(2) (isospin) rotation. Then labeling these doublets by $i = 1,2$ we verify that [23]

$$\begin{aligned} J_\mu^{\mathrm{I}} &= \sum_{i,j=1}^{2} \delta_{ij}\bar{u}_i\gamma_\mu(1+\gamma_5)d_j \\ J_\mu^{\mathrm{II}} &= \sum_{i,j=1} \varepsilon_{ij}\bar{u}_i\gamma_\mu(1+\gamma_5)d_j \end{aligned} \tag{4.47}$$

transform as in eqs. (4.45) and (4.46), respectively. However, the existence of *both* types of quarks within observed particles would lead to a violation of the observed

$$R = \frac{\sigma(e^+e^- \to \text{hadrons})}{\sigma(e^+e^- \to \mu^+\mu^-)} \tag{4.48}$$

ratio discussed earlier. We shall not then discuss a specific dynamical representation of such anomalous currents further but only their phenomenological implications. (Indeed, experimentalists should never be dissuaded by theorists' assertions that certain possibilities are unlikely. Physics is an *experimental* science.)

Weinberg has termed currents with the standard quark model transformation property "first class" while relegating objects transforming as in eq. (4.46) to the "second class" [24]. It then becomes an experimentally interesting challenge to place limits on the possible presence of these second-class currents. Nuclear beta decay is an important laboratory in this regard. When no second class currents are present [25],

(1) If parent-daughter states are isotopic analogs, then

$$\langle I,I_z-1;p_2|J_\mu{}^W|I,I_z;p_1\rangle = \langle I,-I_z;p_1|J_\mu{}^W|I,-I_z+1;p_z\rangle^*$$

yielding (4.49)

$$e = d = j_2 = 0.$$

(2) For mirror β^-,β^+ transitions, none of the structure functions is required to vanish. However, we must have

$$\begin{aligned} &\langle I=0, I_3=0; p_2|J_\mu{}^W|I=1, I_3=1; p_1\rangle \\ &\quad = \langle I=0, I_3=0; p_2|J_\mu^{W\dagger}|I=1, I_3=-1; p_1\rangle \end{aligned} \tag{4.50}$$

yielding

$$a^-,b^-,\ldots h^- = a^+,b^+,\ldots h^+.$$

Then, for example, comparing ft values for mirror electron and positron branches, we require

$$\frac{ft^+}{ft^-} = \left(\frac{c_- + \frac{2}{3}\frac{E_0{}^-}{M} d_-}{c_+ + \frac{2}{3}\frac{E_0{}^+}{M} d_+} \right)^2 \tag{4.51}$$

$$\approx \left(\frac{c_-}{c_+}\right)^2 \left[1 + \frac{2}{3M}\left((E_0{}^- - E_0{}^+)\frac{d^{\mathrm{I}}}{c} + (E_0{}^- + E_0{}^+)\frac{d^{\mathrm{II}}}{c}\right)\right],$$

where we have decomposed the axial tensor d into first-class, second-class pieces $d^{\mathrm{I}}, d^{\mathrm{II}}$, respectively. We could in principle seek evidence for a non-zero d^{II} by looking for dependence of the mirror decay ratios ft^+/ft^- on $E_0{}^- + E_0{}^+$—and such a program has been carried out [26]. In fact, such an analysis performed in 1970 suggested the presence of a significant second class effect [27] and led to a series of experiments described below, which finally settled this issue. The difficulty here is that the size of the expected effects is small—

$$\frac{E_0}{M_N} \sim \frac{10 \text{ MeV}}{1 \text{ GeV}} \sim 1\% \tag{4.52}$$

for $d^{\mathrm{II}}/Ac \sim 0(1)$. In addition, electromagnetic effects can easily produce $(c_-/c_+)^2$ differing from unity by similar amounts. Thus in so-called impulse approximation, which assumes that the nuclear transition takes place as the superposition of single nucleon processes, we find

$$c = g_A \int d^3r_1 \ldots d^3r_A \psi_f{}^*(\mathbf{r}_1, \ldots \mathbf{r}_A) \sum_{i=1}^{A} \tau_i \boldsymbol{\sigma}_i \psi_i(\mathbf{r}_1, \ldots \mathbf{r}_A). \tag{4.53}$$

We then expect $c_+ \neq c_-$ because of wave function differences due to the simple observation that Coulombic effects bind the "last" proton of the β^+ parent nucleus less strongly than the "last" neutron of the β^- parent. This "overlap" asymmetry is quite model-dependent and must be subtracted in order to probe for a bona fide second-class effect [28]. As a result, this method has been essentially abandoned—the systematics are all but impossible to disentangle.

Instead, the search for second-class currents has utilized experiments wherein one looks for a correlation between the outgoing beta particle

and either an initial state polarization/alignment or the direction of a delayed alpha or gamma particle [29]. This procedure has been employed in three systems—$A = 8$, 12, and 20.
$A = 8$: β-α correlation

$$\left.\begin{array}{l} {}^{8}\mathrm{Li} \\ {}^{8}\mathrm{B} \end{array}\right\} \rightarrow {}^{8}\mathrm{Be}^*(2.90\ \mathrm{MeV},\ 2^+) + \begin{array}{l} e^- + \bar{\nu}_e \\ e^+ + \nu_e \end{array}$$
$$\hookrightarrow \alpha + \alpha$$

$A = 12$: β-alignment correlation

$$\left.\begin{array}{l} {}^{12}\mathrm{B} \\ {}^{12}\mathrm{N} \end{array}\right\} \rightarrow {}^{12}\mathrm{C} + \begin{array}{l} e^- + \bar{\nu}_e \\ e^+ + \nu_e \end{array}$$

$A = 20$: β-γ correlation

$$\left.\begin{array}{l} {}^{20}\mathrm{F} \\ {}^{20}\mathrm{Na} \end{array}\right\} \rightarrow {}^{20}\mathrm{Ne}^*(1.63\ \mathrm{MeV},2^+) + \begin{array}{l} e^- + \bar{\nu}_e \\ e^+ + \nu_e \end{array}.$$
$$\hookrightarrow {}^{20}\mathrm{Ne} + \gamma$$

In the $A = 8$ system the combined beta/alpha spectrum from the decay of an unpolarized parent has the form

$$\frac{d\Gamma}{dE_e\, d\Omega_e\, d\Omega_\alpha} \propto 1 + \delta(E)\left(\left(\frac{\mathbf{p}_e \cdot \hat{p}_\alpha}{E_e}\right)^2 - \frac{1}{3}\frac{p_e^2}{E_e^2}\right) - 2\frac{E_e}{Mv^*} \cdot \left(\frac{\mathbf{p}_e}{E_\alpha} \cdot \hat{p}_\alpha\right), \tag{4.54}$$

where v^* is the velocity of the alpha particle in the daughter center of mass frame; for $A = 20$ the beta/photon spectrum for unpolarized decay is given by

$$\frac{d\Gamma}{dE_e\, d\Omega_e\, d\Omega_\gamma} \propto 1 + \frac{1}{2}\delta(E)\left(\left(\frac{p_e \cdot \hat{p}_\gamma}{E_e}\right)^2 - \frac{1}{3}\frac{p_e^2}{E_e^2}\right). \tag{4.55}$$

Experimentally we look for the dependence on the relative angle between the emitted beta particle and the delayed α or γ. In the $A = 12$ case, there exists no delayed emission. Instead, we align the spin of the parent along some direction $\hat{n}$ and measure the dependence of the beta-decay angle with respect to this alignment axis. The spectrum then has the form

$$\frac{d\Gamma}{dE_e d\Omega_e} \propto 1 - 2\Lambda\delta(E)\left(\left(\frac{\mathbf{p}_e \cdot \hat{n}}{E_e}\right)^2 - \frac{1}{3}\frac{p_e^2}{E_e^2}\right), \tag{4.56}$$

with

$$\Lambda = 1 - \frac{3}{2}\langle J_z^2 \rangle \tag{4.57}$$

being the alignment parameter, which measures the deviation of $\langle J_z^2 \rangle$ from the value $\frac{2}{3}$ which would obtain in the unpolarized situation. In each case, the correlation depends on a dynamical component,

$$\delta^{\pm}(E) = \frac{E}{2M}\left(\pm\frac{b + d^{\mathrm{II}}}{c} - \frac{d^{\mathrm{I}}}{c}\right), \tag{4.58}$$

times a geometrical ("Clebsch-Gordon") factor of order unity. These are difficult experiments, since

$$\frac{E}{m_N} \sim 1\%. \tag{4.59}$$

However, they have been undertaken by a number of groups around the world. The analysis for second-class effects is performed by using the difference of the β^-,β^+ correlations in order to remove effects of any first-class induced tensor d^{I}:

$$\delta^-(E) - \delta^+(E) = -\frac{E}{M}\frac{b + d^{\mathrm{II}}}{c}. \tag{4.60}$$

Here, the Gamow-Teller coupling c can be determined from the measured ft value for the transition in question, while the weak magnetism term b can be predicted, using the CVC hypothesis, from the experimental M1 width of the electromagnetic decay of the isotopic analog state, as we saw earlier in our analysis of the shape factor in the $A = 12$ system. The results of these experimental programs are as follows:

	d^{II}/Ac	d^{II}/b	
$A = 8$ [30]	-1.1 ± 0.7	-0.11 ± 0.07	
$A = 12$ [31]	-0.2 ± 0.5	-0.05 ± 0.13	(4.61)
$A = 20$ [32]	0.2 ± 0.8	0.05 ± 0.20	

A similar experiment has been performed in the $A = 19$ system [33]:

$$^{19}\text{Ne} \rightarrow {}^{19}\text{F} + e^+ + \nu_e,$$

wherein one measures the correlation of the outgoing β^+ momentum with the polarization of the initial ^{19}Ne nucleus—$\mathbf{J} \cdot \hat{p}_\beta$. The theoretical expression for this correlation is

$$A(E) = u + v\frac{E}{M}, \tag{4.62}$$

with

$$\frac{v}{u} = -\frac{1}{3}\left(\frac{\frac{2}{\sqrt{3}}a(b+d) + \frac{2}{3}c(5b-d)}{\frac{2}{\sqrt{3}}ac + \frac{2}{3}c^2} - \frac{4ab}{a^2+c^2}\right). \tag{4.63}$$

In this case, since ^{19}F, ^{19}Ne form an isotopic doublet, the axial tensor d must *vanish* in the absence of second-class currents—eq. (4.49). Again the Gamow-Teller can be found from the measured ^{19}Ne ft value, while the Fermi, weak magnetism couplings a and b can be determined from the assumed validity of the CVC hypothesis,

$$\begin{aligned} b &= \frac{1}{\sqrt{3}}(\mu_{\text{Ne}} - \mu_F) \\ a &= \frac{1}{e}(Z_{\text{Ne}} - Z_F), \end{aligned} \tag{4.64}$$

where here Z,μ is the charge, magnetic moment of the corresponding nucleus. By comparing the measured slope with that predicted in the absence of second-class currents—$d = 0$—one can probe for the presence of second-class effects. The experimental result of this measurement is [34]

$$\begin{array}{ccc} A = 19 & d^{\text{II}}/Ac & d^{\text{II}}/b. \\ & 4 \pm 2 & 1 \pm 0.5 \end{array} \tag{4.65}$$

Comparing the numbers quoted in eqs. (4.61) and (4.65), with the impulse approximation prediction

$$\frac{b}{Ac} \simeq \frac{g_M}{g_A} + \frac{g_V}{g_A}\frac{\left\langle f\left\|\sum_i \tau_i \mathbf{L}_i\right\| i\right\rangle}{\left\langle f\left\|\sum_i \tau_i \boldsymbol{\sigma}_i\right\| i\right\rangle} \tag{4.66}$$

we conclude that there is no evidence for the presence of second-class currents at the level of about 20% of weak magnetism. This is, of course, consistent with the quark model expectation. However, the quark model also requires that if one probes with a bit more "magnification," effects should eventually be observed arising from the *u,d* quark mass difference. This is not truly a "second-class" current but can be characterized in a relativistic quark picture by an effective d^{II} coupling [35],

$$\frac{\text{“}d^{\mathrm{II}}\text{”}}{b} \sim \frac{\int d^3r\, r(u_u l_d - u_d l_u)}{\int d^3r\, r(u_u l_d + u_d l_u)} \sim -0.05, \tag{4.67}$$

where u and l are the upper and lower components of *u,d* quark wave functions, as indicated. Probing for effects at this level of precision will require an entirely new generation of experiments. However, it should be of considerable interest since a nonzero effect *is* expected, which can reveal information about the basic quark structure of the weak current.

Finally, let's ask if there is anything to be learned if one *adds* rather than *subtracts* the β^-,β^+ correlation coefficients:

$$\delta^-(E) + \delta^+(E) = -\frac{E}{M}\frac{d^{\mathrm{I}}}{c} \tag{4.68}$$

Obviously, we find in this case the tensor form factor which is "induced" by strong interaction effects from the ordinary first-class axial vector current. This axial tensor has no importance from the point of view of probing for second class currents but *is* an important source of information on the possible relevance of meson exchange effects in nuclear beta decay [36]. We shall examine this aspect in Chapter 7.

4.3 Right-Handed Currents

We have heretofore discussed two "classic" tests of the symmetries of the charged weak current—the validity of the conserved vector current hypothesis and the absence of second-class currents. Both issues are basically settled, and both are in agreement with the "standard" quark-model-based picture of the weak interaction. As emphasized earlier, in such a picture it is difficult to violate these concepts. In fact, both may be considered as tests of the quark model rather than of the symmetries themselves. We now turn, however, to a different issue—that of the "handedness" of the charged weak current. This is an ongoing area of research, and one where the answer is *not* an obvious consequence of the model.

The Feynman–Gell-Mann proposal, as discussed above, is that parity is violated maximally, and experiment confirmed this suggestion—that the weak current responsible for nuclear beta decay involves $(1 + \gamma_5)$ and is purely left-handed. Equivalently, one can say that this feature results because the electroweak interaction involves the local gauge symmetry,

$$SU(2)_L \otimes U(1),$$

requiring the existence of two charged W bosons that couple to left-handed currents ($W_\mu{}^{\pm}$), and a neutral Z boson ($Z_\mu{}^0$) that couples to a linear combination of a neutral left-handed current and the electromagnetic current.

However, this hypothesis raises some interesting philosophical questions. Why should God be left-handed—why should nature choose a left-handed weak interaction instead of say the more symmetric gauge group,

$$SU(2)_L \otimes SU(2)_R \otimes U(1).$$

Such a theory would involve two pairs of charged bosons—one pair,

$$W^{\pm}_{L_\mu},$$

coupling left-handedly, and a second pair,

$$W^{\pm}_{R_\mu},$$

coupling via right-handed currents—and, in addition, a pair of neutral gauge bosons,

$$(Z^0_{L_\mu}, Z^0_{R_\mu})$$

with their own distinctive couplings. Of course, while this scenario is esthetically appealing, in that it has a more symmetric form than the "standard" electroweak model, it is only consistent with low-energy weak phenomenology if the right-handed W-boson is significantly more massive than its left-handed counterpart,

$$M_{W_R} \gg M_{W_L}.$$

The left-handed Fermi constant would then be much greater than the right-handed coupling,

$$G_L \gg G_R = G_L \frac{M^2_{W_L}}{M^2_{W_R}}, \tag{4.69}$$

as observed experimentally. Nevertheless, if we were to perform a *very* high energy experiment,

$$q^2 \gg M_{W_L}^2, M_{W_R}^2,$$

then the effective interaction

$$\mathscr{H}_{\text{int}} \sim g^2 \left\{ \frac{1}{M_{W_L}^2 - q^2} \gamma_\mu(1 + \gamma_5) \otimes \gamma^\mu(1 + \gamma_5) + \frac{1}{M_{W_R}^2 - q^2} \gamma_\mu(1 - \gamma_5) \otimes \gamma^\mu(1 - \gamma_5) \right\} \tag{4.70a}$$

reduces to

$$\mathscr{H}_{\text{int}} \underset{q^2 \gg M_W{}^2}{\widetilde{\quad\quad}} -\frac{g^2}{q^2} \{\gamma_\mu(1 + \gamma_5) \otimes \gamma^\mu(1 + \gamma_5) + \gamma_\mu(1 - \gamma_5) \otimes \gamma^\mu(1 - \gamma_5)\}$$
$$= -2\frac{g^2}{q^2}(\gamma_\mu \otimes \gamma^\mu + \gamma_\mu\gamma_5 \otimes \gamma^\mu\gamma_5), \tag{4.70b}$$

which is parity conserving—no handedness remains! This left-right symmetric idea has such an appealing form that many theorists have analyzed such models [37] and numerous experiments have probed for the existence of small right-handed couplings in weak interactions.

Before we review some of these experiments, we note one slight modification of our previous simplified discussion. As in the case of the weak couplings of quark fields discussed earlier, there is no reason here for the gauge bosons having purely left- and right-handed weak couplings ($W_L{}^\pm, W_R{}^\pm$) to be identical to mass eigenstates of the strong interaction (W_1, W_2). In general, such mass eigenstates are represented in terms of a mixture of the chirality eigenstates,

$$\begin{aligned} W_1 &= \cos\chi W_L - \sin\chi W_R \\ W_2 &= \sin\chi W_L + \cos\chi W_R \end{aligned} \quad \text{with mass} \quad \begin{aligned} &m_1 \\ &m_2, \end{aligned} \tag{4.71}$$

where χ is an unknown mixing angle. We also define the parameter

$$\sigma = \frac{m_1{}^2}{m_2{}^2}, \tag{4.72}$$

which measures the relative scale of right/left-handed couplings. The standard model is then recovered in the limit $\sigma, \chi \to 0$. However, in the more general case, with both σ, χ nonzero but small, we can write the effective semileptonic weak interaction as [38]

$$\mathscr{H}_w \sim \frac{G}{\sqrt{2}}[\gamma_\mu(1+\gamma_5)\otimes\gamma^\mu(1+\varepsilon\gamma_5)+\gamma_\mu(1-\gamma_5)\otimes\gamma^\mu(x-y\varepsilon\gamma_5)], \quad (4.73)$$

where the coupling on the left, right-hand side of the $\otimes$ symbol is associated with the lepton, hadron sectors, respectively. Here we have defined

$$x \approx \sigma - \chi, \qquad y = \sigma + \chi, \qquad \varepsilon = \frac{1-x}{1-y}. \quad (4.74)$$

Next we consider the status of experiments that can place restrictive limits on σ,χ. We shall discuss three examples and indicate areas where future activity is planned.

Electron/Positron Helicity Studies

As mentioned earlier, the existence of the purely left-handed coupling $(1+\gamma_5)$ implies that the polarization of a relativistic β^- particle must be opposite to its momentum; that is, the longitudinal polarization is given simply by

$$P_L = \frac{v}{c}. \quad (4.75)$$

On the other hand, the presence of right-handed currents will modify this prediction. In the case of a pure Gamow-Teller transition we find

$$P_L = \frac{v}{c}\frac{1-y^2}{1+y^2}. \quad (4.76)$$

The statistical average of a number of such measurements [39]

$$P_L^{\text{exp}} = \frac{v}{c}(0.992 \pm 0.005) \quad (4.77)$$

then provides a limit on the combination $\sigma + \chi$

$$|y| = |\sigma + \chi| < 0.10 \quad \text{at } 90\% \text{ C.L.} \quad (4.78)$$

A number of recent experiments, instead of attempting to measure the *absolute* longitudinal polarization, have looked instead at the *ratio* of positron polarizations emitted in Fermi vs. Gamow-Teller decays, for which the presence of right-handed currents would give

$$P_L{}^F/P_L^{GT} \cong (1 - 2x^2 + 2y^2) = (1 + 8\sigma\chi). \tag{4.79}$$

A recent experiment by the Gröningen group utilizing a fourfold Bhabbha polarimeter determined [40]

$$P_L{}^F/P_L^{GT} = 1.003 \pm 0.004, \tag{4.80}$$

comparing the Fermi decay of $^{26}\mathrm{Al}^{\mathrm{m}}$ with the Gamow-Teller transition $^{30}\mathrm{P} \to {}^{30}\mathrm{Si} + e^+ + \nu_e$. (The nearly identical Z-values, 13 vs. 15, and end-point energies, 3211 KeV vs. 3205 KeV, for these processes are useful in canceling Coulomb and radiative corrections.) Thus we have from the Gröningen experiment

$$\sigma\chi = (0.4 \pm 0.5) \times 10^{-3}. \tag{4.81}$$

Two other groups have been employing time-resolved positronium-annihilation spectroscopy as a probe of the relative longitudinal polarization [41]. A Michigan-Toronto collaboration has achieved a comparison between the Fermi/Gamow-Teller decay of $^{25}\mathrm{Al}$ and the $^{26}\mathrm{Al}^{\mathrm{m}}$ Fermi transition at the 0.005 level while a group at Louvain-la-Neuve has achieved 0.003 in a comparison of the decays of $^{10}\mathrm{C}$ (Gamow-Teller) and $^{14}\mathrm{O}$ (Fermi) [42]. Both results are preliminary.

Correlation Measurements in Beta-Decay

A second technique that has been employed to probe for right-handed currents involves the comparison of the $\mathbf{J} \cdot \hat{p}_\beta$ correlation and ft value in mixed Fermi-Gamow-Teller decays. The most sensitive such test involves the positron decay of polarized $^{19}\mathrm{Ne}$,

$$^{19}\mathrm{Ne} \to {}^{19}\mathrm{F} + e^+ + \nu_e,$$

for which the decay spectrum has the form

$$\frac{d\Gamma_\beta}{d\Omega_e} \sim 1 + AP\hat{\mathbf{J}} \cdot \mathbf{p}_\beta/E_\beta + \ldots . \tag{4.82}$$

Here $P\hat{\mathbf{J}}$ is the parent nucleus polarization, and

$$A = \frac{\dfrac{2}{\sqrt{3}}c\left(a + \dfrac{c}{\sqrt{3}}\right) - 2y\dfrac{c}{\sqrt{3}}\left(xa + y\dfrac{c}{\sqrt{8}}\right) + T_1}{a^2 + c^2 + x^2a^2 + y^2c^2 + T_2} \tag{4.83}$$

is the asymmetry coefficient, with T_1, T_2 being known small ($O[1\%]$) corrections due to recoil order structure constants. A second constraint comes from the measured *ft* value for the transition, which may be compared to the corresponding number found in analysis of $0^+ - 0^+$ Fermi decays.

$$\frac{ft^{\text{Fermi}}}{ft^{^{19}\text{Ne}}} = \frac{a^2 + c^2 + x^2 a^2 + y^2 c^2 + T_3}{a_F{}^2(1 + x^2)}, \tag{4.84}$$

where again T_3 is a known $0(1\%)$ correction. We can then use the experimental values of A and *ft* to set limits in the two parameter σ,χ space [43]. The results are shown in Figure 4.3 and are seen to be complementary to the bounds found in the helicity experiments discussed above.

The reasons behind the use of ^{19}Ne are twofold. First, atomic beam techniques allow a virtually 100% polarized sample of ^{19}Ne atoms to be

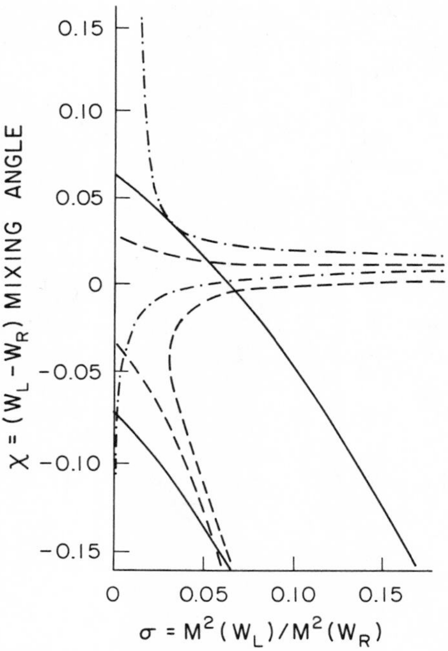

FIGURE 4.3. Plotted are the experimental restrictions in χ,σ space arising from direct helicity measurements in Gamow-Teller decays (solid), from a comparison of Gamow-Teller and Fermi helicities (dot-dash) and from the asymmetry parameter in ^{19}Ne (dash) [49].

prepared [44]. Second, the measured ft value,

$$ft^{^{19}\mathrm{Ne}} \cong 1729 \text{ sec}, \tag{4.85}$$

together with the shell model result that

$$\frac{c}{a} \simeq \frac{g_A}{g_V} \frac{\langle {}^{19}\mathrm{F} | \sum \tau_i \boldsymbol{\sigma}_i | {}^{19}\mathrm{Ne} \rangle}{\left\langle {}^{19}\mathrm{F} \middle| \sum_i \tau_i \middle| {}^{19}\mathrm{Ne} \right\rangle} < 0 \tag{4.86}$$

and the number measured in pure Fermi transitions

$$ft^{\mathrm{Fermi}} \cong 3072 \text{ sec}, \tag{4.87}$$

yields

$$\left(\frac{c}{a}\right)^{^{19}\mathrm{Ne}} \cong -\left(\frac{2ft^{\mathrm{Fermi}}}{ft^{^{19}\mathrm{Ne}}} - 1\right)^{1/2} \simeq -1.60, \tag{4.88}$$

which is very close to the value

$$\left(\frac{c}{a}\right)^{^{19}\mathrm{Ne}} = -\sqrt{3} \tag{4.89}$$

for which the asymmetry would vanish in the absence of right-handed currents. The asymmetry predicted using the experimental value of the Gamow-Teller, eq. (4.49),

$$A_{\mathrm{left}}^{\mathrm{theo}} \simeq \frac{2}{\sqrt{3}} \frac{c}{a} \frac{\left(1 + \dfrac{1}{\sqrt{3}} \dfrac{c}{a}\right)}{1 + \dfrac{c^2}{a^2}} \simeq -0.04, \tag{4.90}$$

is then quite small and is consequently very sensitive to any right-handed effects that might be present.

Use of the asymmetry and ft value in tandem is also be possible for the case of neutron beta decay, for which recent experiments at Grenoble have yielded the very precise value [45]

$$A = -0.1146 \pm 0.0019. \tag{4.91}$$

Unfortunately, the ft value is not yet known accurately enough to enable a study of possible right-handed currents to be competitive with the ^{19}Ne analysis [46].

Muon-Decay Spectrum

A third way of limiting possible right-handed structures is from the purely leptonic sector via careful analysis of the electron spectrum resulting from the decay of polarized muons. In the presence of right-handed currents, the decay spectrum can be written as [47]

$$\begin{aligned}\frac{d^2Q}{x^2\,dx\,d(\cos\theta)} &\sim 3 - 2x + \left(\frac{4}{3}\rho - 1\right)(4x - 3) + 12\frac{m_e}{xm_\mu}(1 - x)\eta \\ &\quad + \xi P_\mu \cos\theta\left[\left(\frac{4}{3}\delta - 1\right)(4x - 3) + 2x - 1\right].\end{aligned} \tag{4.92}$$

Here $x \equiv E_e/E_{\max}$ and the so-called Michel parameters ρ, η, ξ, δ are given by

$$\begin{aligned}\rho &\cong \frac{3}{4}\left(1 - \frac{1}{2}(x - y)^2\right) \\ \xi &\cong 1 - x^2 - y^2 \\ \delta &\approx \frac{3}{4} \\ \eta &\approx 0.\end{aligned} \tag{4.93}$$

A group at TRIUMF has recently announced very precise results from such an experiment. Using either direct analysis of the decay spectrum or by employing the technique of muon spin rotation they obtain comparable limits

$$\begin{aligned}&e^+ \text{ spectrum: } \frac{\xi P_\mu \delta}{\rho} > 0.9959 \quad \text{at } 90\%\ \text{C.L. [48]} \\ &\mu sr\text{:} \qquad\qquad \frac{\xi P_\mu \delta}{\rho} > 0.9955 \quad \text{at } 90\%\ \text{C.L. [49]}\end{aligned} \tag{4.94}$$

Connection with right-handed parameter space is provided by the expression

$$\frac{\xi P_\mu \delta}{\rho} \cong 1 - 2(\sigma + \chi)^2 - 2\sigma^2, \tag{4.95}$$

which yields the constraints indicated in Figure 4.4.

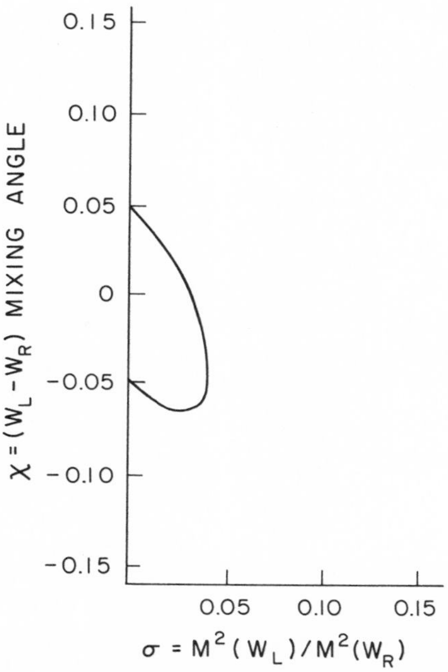

FIGURE 4.4. Plotted are the experimental restrictions in χ,σ space arising from measurement of $\xi P_\mu \delta/\rho$ in muon decay [49].

An SIN collaboration has recently announced results of a new type of experiment sensitive to $P_\mu \xi$, yielding $P_\mu \xi = 1.0027 \pm 0.0084$ [50]. However, the limits obtained thereby are so far not as strong as those found by other means.

A curious feature here is the surprisingly weak—$\sim 5\%$—bounds on right-handed amplitudes σ,χ that have been placed despite incredibly precise—$\sim 0.1\%$—experiments. This is because left- and right-handed sectors do not interfere with one another. Right-handed effects always arise as the square of σ,χ or as the product $\sigma\chi$. Thus, a 0.1% experiment only provides a limit $\sim \sqrt{0.001} \approx 3\%$!

Having summarized the impact of these careful experiments on our knowledge of possible right-handed currents, I wish now to put in a good word for the efforts of theorists who, provided certain assumptions are made about the validity of their models, are able to limit things even further. I will resist the temptation to give any details of these calculations. However, it should be noted that all such models assume left- and right-handed KM angles to be identical, which is natural, but not required in left-right gauge models [51]. Then,

(1) Analysis of semileptonic weak processes provides bounds on elements of the KM matrix. If we make the assumption that there exist only three generations, then the required unitarity of U_{KM} can be exploited in order to give a limit [52]

$$|\chi| < 0.005 \tag{4.96}$$

on possible left-right mixing.

(2) Both amplitude and slope of the various $K \to 3\pi$ decays are well predicted in terms of known (experimental) $K \to 2\pi$ transition rates [53] if we utilize the assumption that the $\Delta S = 1$ weak nonleptonic Hamiltonian is constructed from purely left-handed currents, in which case

$$[F_a{}^5, H_w(\Delta S = 1)] = [F_a, H_w(\Delta S = 1)] \qquad a = 1,2,3, \tag{4.97}$$

where

$$F_a{}^5 = \int d^3x A_0{}^a(\mathbf{x},t) \quad \text{and} \quad F_a = \int d^3x V_0{}^a(\mathbf{x},t) \tag{4.98}$$

are the axial and polar vector charges, respectively. The presence of right-handed currents will alter this commutator identity, and the agreement of the experimental $K \to 3\pi$ amplitude with the predictions based on eq. (4.97) can be used as a probe for possible right-handed effects. In this way we find the mixing angle limit $|\chi| < 0.004$ [54].

(3) The $K_L - K_S$ mass difference in left-right models is not subject to the usual helicity arguments that suppress calculations of Δm in purely left-handed models [55], enabling a strong limit to be placed upon the mass of the right-handed W boson [56],

$$m_{W_R} < 1.6 \text{ TeV} \tag{4.99}$$

or

$$\sigma < 0.003. \tag{4.100}$$

Although these limits obtained from theoretical analysis are impressive and about a factor of ten better than those obtained experimentally, we should remember that the more precise values are obtained only at the cost of additional model dependence. Thus, experimentalists (and theorists) should not be discouraged from attempting to continue their work in bounding the size of right-handed couplings.

4.4 Chiral Invariance and PCAC

The last property of the weak current that we shall discuss in this chapter is that of chiral symmetry [57], which—though not an exact invariance—plays an important role in our understanding of particle interactions. The basic idea here is to express the Lagrangian for *u*,*d* quarks (which is the piece relevant for nuclear physics),

$$\mathscr{L} = \bar{u}(i\not{\partial} - m_u)u + \bar{d}(i\not{\partial} - m_d)d + \dots \tag{4.101}$$

in terms of its left- and right-handed components using the projection operators $P_{L,R} = \frac{1}{2}(1 \pm \gamma_5)$

$$\begin{aligned} \mathscr{L} = {} & \bar{u}_L i\not{\partial} u_L + \bar{u}_R i\not{\partial} u_R + \bar{d}_L i\not{\partial} d_L + \bar{d}_R i\not{\partial} d_R \\ & - m_u(\bar{u}_L u_R + \bar{u}_R u_L) - m_d(\bar{d}_L d_R + \bar{d}_R d_L) + \dots . \end{aligned} \tag{4.102}$$

We see that the "energy" terms $i\not{\partial}$ contain only $\bar{q}q$ mixtures of identical handedness, whereas the existence of mass allows mixing between left- and right-handed quark components. In the (mathematical) limit $m_u = m_d = 0$, the Lagrangian is invariant under global transformations in either the left- or right-handed sectors:

$$\begin{aligned} \psi_L &\to \exp(-i\boldsymbol{\alpha} \cdot \boldsymbol{\tau}/2)\psi_L \\ \psi_R &\to \exp(-i\boldsymbol{\beta} \cdot \boldsymbol{\tau}/2)\psi_R \end{aligned} \quad \text{with } \psi = \begin{pmatrix} u \\ d \end{pmatrix}. \tag{4.103}$$

Thus, a global $SU(2)_L \otimes SU(2)_R$ symmetry exists which is usually called "chiral invariance," meaning that under such transformations the handedness does not change. Obviously, this is an *exact* symmetry only if the *u*,*d* quarks are massless. However, since $m_u, m_d \sim 10$ MeV $\ll$ a typical hadronic energy scale, approximate chiral symmetry should also be a useful concept. In order to see how the physics is realized in this case, we recall the idea of spontaneous symmetry breaking wherein the ground state or "vacuum" is not invariant under a symmetry transformation of the Lagrangian. Then, if Q is a would-be total charge operator, spontaneous symmetry breaking requires

$$Q|0\rangle \neq 0 \quad \text{and} \quad \langle 0|Q^2|0\rangle \neq 0. \tag{4.104}$$

The Goldstone theorem [58] states that when a continuous symmetry is broken in this fashion, massless particles, which carry the quantum numbers of the charge Q, are generated. In the case at hand, we assume that the chiral symmetry is broken spontaneously with

$$Q_5|0\rangle \neq 0, \tag{4.105}$$

where

$$Q_5{}^{\mathrm{i}} = \int d^3x A_i{}^0(\mathbf{r},t) \qquad i = 1,2,\ldots,8 \tag{4.106}$$

is the axial charge. We expect then the presence of massless pseudoscalar "Goldstone bosons." Since the only known massless particles are the photon and neutrino, neither of which is a pseudoscalar, something doesn't seem to work. However, recall that the Goldstone theorem is derived by assuming that the Lagrangian is chiral-symmetric. In the real world, chiral symmetry is broken in the Lagrangian by the mass terms m_u,m_d. Thus, we might anticipate the existence of light (but not massless) pseudoscalar particles with masses proportional to the current quark masses m_u, m_d,m_s. Experimentally, taking $m_N \sim 1$ GeV as a typical hadronic mass scale, we find

$$\frac{m_\pi{}^2}{m_N{}^2} \sim 0.02, \qquad \frac{m_K{}^2}{m_N{}^2} \sim 0.25, \qquad \frac{m_\eta{}^2}{m_N{}^2} = 0.30, \tag{4.107}$$

so these expectations are borne out—the pion is a Goldstone boson with its mass given by the matrix element

$$m_\pi{}^2 = \langle\pi|m_u\bar{u}u + m_d\bar{d}d|\pi\rangle. \tag{4.108}$$

Now consider the limit in which m_u,m_d are set equal to zero, so that the pion mass vanishes. If we look at neutron beta decay via the axial current in this limit, we find the matrix element

$$\langle p|A_\mu|n\rangle = \bar{u}(p)\left(g_A\gamma_\mu\gamma_5 - g_P\frac{1}{2M}q_\mu\gamma_5\right)u(n), \tag{4.109}$$

where $g_A = 1.25$ is the usual axial coupling and g_P is the induced pseudoscalar. Because of the existence of the pion a pole occurs in the induced term—the neutron can decay virtually to a proton and a (massless) pion, which is then annihilated to the vacuum through the axial current (cf. Figure 4.5)

$$\langle p|A_\mu|n\rangle_{\text{pole}} = \bar{u}(p)\gamma_5 u(n) g_{\pi np}\frac{1}{q^2}\sqrt{2}F_\pi q_\mu, \tag{4.110}$$

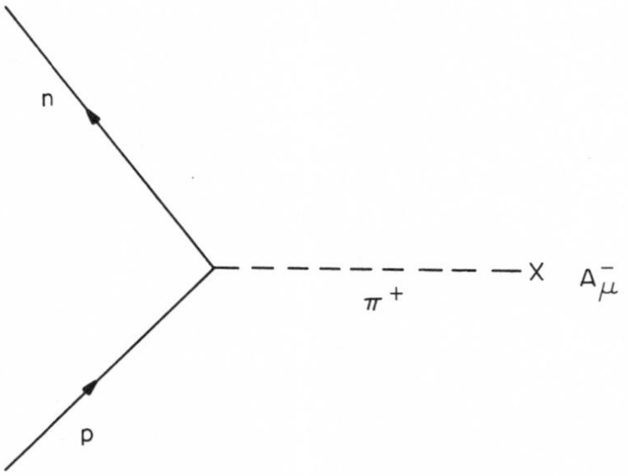

FIGURE 4.5. The pion pole contribution to the axial current matrix element of the nucleon.

where $g_{\pi np}$ is the usual pseudoscalar coupling constant

$$g_{\pi np} = \sqrt{2} g_r \quad \text{with} \; \frac{g_r^2}{4\pi} \simeq 14, \tag{4.111}$$

and F_π is measured in pion beta decay

$$\langle 0|A_\mu|\pi^+\rangle = \sqrt{2} F_\pi q_\mu \quad \text{with } F_\pi \simeq 94 \text{ MeV}. \tag{4.112}$$

For $m_u, m_d = 0$ the Dirac equation becomes

$$i\not{\partial} u = i\not{\partial} d = 0, \tag{4.113}$$

resulting in a divergenceless axial current

$$i\partial_\mu A^\mu = i\partial_\mu \bar{u}\gamma^\mu\gamma_5 d = 0. \tag{4.114}$$

If we take a matrix element of this relation and recall that the divergence is equivalent to contraction with the momentum transfer, we find

$$\begin{aligned} 0 = \langle p|i\partial_\mu A^\mu|n\rangle &= \langle p|g_A \not{q}\gamma_5 - g_P \frac{1}{2M} q^2 \gamma_5|n\rangle \\ &= \text{(using the Dirac equation)} \; \langle p|\left(-2M g_A - g_P \frac{1}{2M} q^2\right)\gamma_5|n\rangle. \end{aligned} \tag{4.115}$$

Finally, we assume that g_P is dominated by the pion pole

$$g_P \approx g_P^{\text{Pole}} = -2M \cdot \frac{\sqrt{2}g_r \cdot \sqrt{2}F_\pi}{q^2}, \tag{4.116}$$

so that

$$\frac{g_A}{F_\pi} = \frac{g_r}{M}, \tag{4.117}$$

which is called the Goldberger-Treiman relation [59], relating the weak interaction parameters g_A and F_π with the strong pion nucleon coupling constant g_r. Experimentally it is well satisfied, since

$$13.3\ \text{GeV}^{-1} = \frac{g_A}{F_\pi} \quad \text{vs.} \quad 14.1\ \text{GeV}^{-1} = \frac{g_r}{M}. \tag{4.118}$$

Also, we note that

$$\langle 0|\partial_\mu A^\mu|\pi\rangle = \sqrt{2}F_\pi q^2 = 0 \quad \text{since } q^2 = m_\pi{}^2 = 0. \tag{4.119}$$

Thus, everything is consistent except that our calculation was performed in an imaginary chiral invariant ($m_u = m_d = 0$) world, whereas the experimental strong, weak couplings are measured in the real ($m_u, m_d \neq 0$) world. However, as we shall see, the Goldberger-Treiman relation remains valid when we allow quark masses to become nonzero.

One thing that is changed in the real world limit is the divergenceless condition on the axial current, since we now have

$$-i\partial_\mu A^\mu = (m_u + m_d)\bar{u}\gamma_5 d. \tag{4.120}$$

Taking a matrix element between a pion and the vacuum, we determine

$$\langle 0|i\partial_\mu A^\mu|\pi\rangle = \sqrt{2}F_\pi q^2 = \sqrt{2}F_\pi m_\pi{}^2. \tag{4.121}$$

Thus, we find

$$m_\pi{}^2 = \frac{m_u + m_d}{\sqrt{2}F_\pi}\langle 0|\bar{u}\gamma_5 d|\pi\rangle, \tag{4.122}$$

which is consistent with the chiral-symmetric idea that the pion would be massless in the limit $m_u, m_d \to 0$.

Since the pion mass is so small, it is sensible to assume that single pion effects dominate matrix elements of $\partial_\mu A^\mu$. We then postulate the partially conserved axial current or PCAC hypothesis [60]

$$\partial_\mu A^\mu \cong \sqrt{2} F_\pi m_\pi^{\ 2} \phi_\pi, \tag{4.123}$$

where ϕ_π is the pion field. The term "partial" conservation just refers to the feature that the right-hand side *would* be zero in the limit $m_u, m_d \to 0$. We can use this assumption—eq. (4.123)—to analyze the axial current matrix element in neutron beta decay, yielding

$$\begin{aligned} \langle p|i\partial_\mu A^\mu|n\rangle &= \langle p|g_A \not{q} \gamma_5 - g_P \frac{1}{2M} q^2 \gamma_5|n\rangle \\ &= \langle p|\left(-2Mg_A - g_P \frac{1}{2M} q^2\right)\gamma_5|n\rangle \\ &= \langle p|\gamma_5|n\rangle \frac{2F_\pi m_\pi^{\ 2} g_r}{q^2 - m_\pi^{\ 2}}. \end{aligned} \tag{4.124}$$

By pion pole dominance we have

$$g_P \cong -\frac{2M \cdot 2F_\pi g_r}{q^2 - m_\pi^{\ 2}}. \tag{4.125}$$

and

$$2Mg_A = \frac{2F_\pi g_r}{q^2 - m_\pi^{\ 2}} (q^2 - m_\pi^{\ 2}) = 2F_\pi g_r, \tag{4.126}$$

so that we recover the Goldberger-Treiman relation, as promised.

We also have a prediction for the value of the induced pseudoscalar—eq. (4.125). Can this be verified? Consider first nuclear beta decay. Although g_P/g_A is large,

$$\frac{g_P}{g_A} = \frac{4MF_\pi g_r}{m_\pi^{\ 2} g_A} \sim 200, \tag{4.127}$$

any induced pseudoscalar effects are strongly suppressed. The point is that since

$$q_\mu \bar{u}(e)\gamma^\mu(1 + \gamma_5)v(\nu) = m_e \bar{u}(e)(1 + \gamma_5)v(\nu), \tag{4.128}$$

any pseudoscalar effects must be of order [61]

$$\left(\frac{m_e{}^2}{2ME_e}\right) \lll 1. \tag{4.129}$$

However, if we consider muon capture, we have

$$q^\mu \bar{u}(\nu)\gamma_\mu(1 + \gamma_5)u(\mu) = m_\mu \bar{u}(\nu)(1 + \gamma_5)u(\mu) \tag{4.130}$$

and effects are

$$\left(\frac{m_\mu}{2M}\right) \sim 5\%. \tag{4.131}$$

Muon capture then offers powerful advantages relative to beta decay in looking at effects of the induced pseudoscalar because of the much larger muon mass. However, we also pay a price. In beta decay, besides the measurement of the decay rate itself, we can also look at energy spectra or at various correlation functions, as noted previously. That is *not* the case for muon capture, for which typically one can measure only a single number—the capture rate Γ_μ. Nevertheless, we *can* learn about the size of the induced pseudoscalar provided we are willing to make a number of (reasonable) assumptions. In terms of the structure functions defined in section 4.1, the capture rate for an allowed transition is found to be [62]

$$\Gamma_\mu \propto a^2 + c^2 + \frac{m_\mu}{M}\left(a^2 + \frac{1}{3}\left(c^2 + cb - ch\frac{m_\mu}{2M} - cd\right)\right). \tag{4.132}$$

However, we must recall that each of these form factors is a function of q^2, which in the case of nuclear beta decay is essentially zero, but which for muon capture takes the value

$$q^2 = (p_\mu - p_\nu)^2 = m_\mu{}^2 - 2m_\mu E_\nu \approx -m_\mu{}^2. \tag{4.133}$$

In order to learn about the induced pseudoscalar term h in eq. (4.132), we must know the other form factors completely. This is usually done as follows:

1. From CVC we have $e = 0$ and $a(q^2)$,$b(q^2)$ can be predicted in terms of electron scattering measurements involving the electromagnetic analog of the muon capture transition.

2. From the validity of the impulse approximation [63]

$$\begin{aligned} c(q^2) &= \left\langle \beta \left\| \sum_i \tau_i{}^{\pm} \boldsymbol{\sigma}_i j_0(qr) \right\| \alpha \right\rangle \\ b(q^2) &= 4.7 \left\langle \beta \left\| \sum_i \tau_i{}^{\pm} \boldsymbol{\sigma}_i j_0(qr) \right\| \alpha \right\rangle \\ &\quad + \left\langle \beta \left\| \sum_i \tau_i{}^{\pm} \mathbf{L}_i \cdot \frac{3 j_1(qr)}{qr} \right\| \alpha \right\rangle \approx 4.7 c(q^2) \end{aligned} \tag{4.134}$$

we assume

$$\frac{c(q^2)}{c(0)} \cong \frac{b(q^2)}{b(0)}, \tag{4.135}$$

where $c(0)$ is taken from the ft value of the analogous beta-decay reaction.

3. Finally, for a capture between isotopic analog states we can use $d = 0$ from the absence of second-class currents. For a nonanalog capture, $d(0)$ is taken from the beta-decay measurements, and the scaling law

$$\frac{d(q^2)}{d(0)} \cong \frac{b(q^2)}{b(0)} \tag{4.136}$$

is used to predict the induced tensor that is relevant for the muon capture reaction [64].

Thus, we have been forced to make a number of assumptions. All are reasonable, but this is the price we must pay in order to learn the value of the induced pseudoscalar. Generally, the results are parameterized in terms of

$$r_P = \frac{h(q^2)}{c(q^2)} \frac{m_\mu}{2M} \approx (\text{via pole dominance}) \frac{2Mm_\mu}{m_\pi{}^2 - q^2} \approx 7.1. \tag{4.137}$$

Experimentally, we find in this way

$$\begin{aligned} r_P^{\text{exp}} &= 7.4 \pm 2.0 \quad \text{for } \mu^- + {}^3\text{He} \to {}^3\text{H} + \nu_\mu \text{ [65]} \\ r_P^{\text{exp}} &= 6.5 \pm 2.4 \quad \text{for } \mu^- + p \to n + \nu_\mu \text{ [66]}, \end{aligned} \tag{4.138}$$

in good agreement with our prediction.

In one case, it is possible to do even better, and that is

$$\mu^- + {}^{12}\text{C} \rightarrow {}^{12}\text{B} + \nu_\mu,$$

for which the decay rate [67]

$$\Gamma_\mu = (6.2 \pm 0.3) \times 10^3 \text{ sec}^{-1} \tag{4.139}$$

has been determined. However, ^{12}B is itself unstable and quickly decays back to ^{12}C from which it started. Analysis of the beta-decay spectrum can then reveal the polarization of the ^{12}B nucleus resulting from the capture reaction, yielding [68]

$$P_{\text{av}}({}^{12}\text{B}) = 0.47 \pm 0.05. \tag{4.140}$$

The nice thing about this measurement is that, since it involves ratios of form factors, it is nearly independent of assumptions about the q^2 dependence of the various structure functions. A third measurement that has been made in this system is the longitudinal polarization (along the decay direction) yielding [69]

$$\frac{P_{\text{av}}({}^{12}\text{B})}{P_L({}^{12}\text{B})} = -0.506 \pm 0.041. \tag{4.141}$$

Analysis of these experimental numbers is able to give a fairly convincing value of the pseudoscalar form factor

$$r_P^{\text{exp}} = 9.1 \pm 1.7, \tag{4.142}$$

again in good agreement with our theoretical expectations—eq. (4.137).

In principle, we can get additional information from muon capture reactions by using a polarized target or the residual polarization of the captured muons [70]. In this case, correlations between the recoil nucleus and either initial state or final state polarization are sensitive to recoil form factors such as the induced pseudoscalar. However, these are difficult experiments and have not yet been performed.

A second promising avenue is that of radiative muon capture,

$$\mu^- + A + A' + \nu_\mu + \gamma.$$

Early experiments were ambiguous and suffered from background problems, the latter being a result of the small ($\sim 10^{-5}$) branching ratio for this process. Finally, about ten years ago a measurement of ^{40}Ca capture solved much of the neutron background problem by converting the gamma rays into electrons [71], yielding

$$\frac{h_{\text{expt}}}{h_{\text{PCAC}}} = 0.95 \pm 0.23, \tag{4.143}$$

in good agreement with the PCAC hypothesis. A second experiment involving ^{40}Ca found a somewhat smaller but not inconsistent value [72]

$$\frac{h_{\text{expt}}}{h_{\text{PCAC}}} = 0.59 \pm 0.22. \tag{4.144}$$

A radiative capture on hydrogen measurement is presently underway at TRIUMF, but it will be a number of years before results are forthcoming.

Finally, recent measurements by a Swiss-Japanese collaboration involving radiative muon capture on a range of nuclear targets—from ^{12}C to ^{209}Bi—appear to indicate a tendency to agree with the simple PCAC prediction in the case of lighter nuclei with a systematic decrease for large A [73]. This result is not unexpected, however, and may actually confirm theoretical expectations concerning the properties of the axial current when embedded within a nuclear medium [74]. A second prediction of such theories—axial quenching, whereby the axial coupling strength g_A within the nucleus is somewhat reduced from its free space value [75]—also has received support from a systematic study of beta-decay rates in S/D shell nuclei [76]. However, a detailed discussion of these interesting phenomena would take us far afield.

Overall then, a series of measurements involving muon capture has basically confirmed the PCAC hypothesis. At the present time experiments involving correlations are also underway and should lead to a significant improvement in our knowledge of the induced pseudoscalar coupling.

Appendix on Chiral Symmetry

Chiral symmetry has implications for nuclear physics far deeper than those associated with the simple PCAC hypothesis which has just been discussed. These can perhaps best be expressed by means of a so-called effective Lagrangian [77], which includes the constraints of chiral invariance in terms of the degrees of freedom—baryons and mesons—actually observed experimentally rather than in terms of "unphysical" and unobservable quarks that appear in the more fundamental QCD Lagrangian.

To see how this is accomplished, we temporarily omit the baryonic degrees of freedom and consider only the pseudoscalar mesons. It is easiest to deal not with the meson fields themselves,

$$\phi_i(x) \qquad i = 1,2,3,$$

but rather with the nonlinear combination,

$$U(x) = \exp\left(\frac{i}{F_\pi}\sum_{j=1}^{3}\tau_j\phi_j(x)\right), \tag{A1}$$

where

$$\tau_1 = \begin{pmatrix} 0 & 1 \\ 1 & 0 \end{pmatrix} \qquad \tau_2 = \begin{pmatrix} 0 & -i \\ i & 0 \end{pmatrix} \qquad \tau_3 = \begin{pmatrix} 1 & 0 \\ 0 & -1 \end{pmatrix} \tag{A2}$$

are the usual Pauli isospin matrices, and $F_\pi \approx 94$ MeV is the pion decay constant discussed above. Under an arbitrary chiral transformation

$$(L,R)\text{: } \mathrm{SU}(2)_L \otimes \mathrm{SU}(2)_R$$

we define

$$U(x) \to LU(x)R^{-1}. \tag{A3}$$

Chirally symmetric interactions of pseudoscalar mesons must then be described in terms of an effective Lagrangian constructed from $U(x)$ and its derivatives, which are invariant under both Lorentz transformations and under an arbitrary global chiral rotation. The simplest such invariant combination is

$$\mathrm{Tr}\, UU^\dagger \xrightarrow[L,R]{} \mathrm{Tr}\, LUR^{-1}RU^\dagger L^\dagger = \mathrm{Tr}\, UU^\dagger. \tag{A4}$$

However, there is no content here since U is unitary

$$UU^\dagger = U^\dagger U = 1. \tag{A5}$$

The next simplest Lorentz invariant and chiral invariant combination involves

$$\frac{1}{4}F_\pi{}^2\, \mathrm{Tr}\, \partial_\mu U \partial^\mu U^\dagger, \tag{A6}$$

where the constant in front is chosen so that the leading term in the expansion

$$\begin{aligned}\mathscr{L} &= \frac{1}{4} F_\pi^{\,2} \operatorname{Tr} \partial_\mu U \partial^\mu U^\dagger \\ &= \frac{1}{2} \partial_\mu \boldsymbol{\phi} \cdot \partial^\mu \boldsymbol{\phi} + \frac{1}{2F_\pi^{\,2}} (\partial_\mu \boldsymbol{\phi} \cdot \boldsymbol{\phi})^2 + \frac{1}{6F_\pi^{\,2}} \boldsymbol{\phi} \cdot \boldsymbol{\phi}\boldsymbol{\phi} \cdot \Box \boldsymbol{\phi} + \ldots\end{aligned} \tag{A7}$$

is simply the kinetic energy term of the pionic Lagrangian. It is not possible to find a corresponding form which is invariant under chiral rotations and which yields a pion mass, but this is as expected from the result that $m_\pi^{\,2} \to 0$ in the chiral limit $m_u, m_d \to 0$. The mass is not a chiral scalar but rather it is associated with a piece of the effective chiral Lagrangian,

$$\operatorname{Tr}(U + U^\dagger - 2), \tag{A8}$$

which transforms under chiral rotations in the same way as the mass term in the QCD Lagrangian,

$$\mathscr{L}_m = m_u \bar{u} u + m_d \bar{d} d. \tag{A9}$$

The coefficient of this term is chosen so that the pion mass term has the correct normalization,

$$\begin{aligned}\mathscr{L} &= \frac{F_\pi^{\,2} m_\pi^{\,2}}{4} \operatorname{Tr}(U + U^\dagger - 2) \\ &= -\frac{1}{2} m_\pi^{\,2} \boldsymbol{\phi} \cdot \boldsymbol{\phi} + \frac{m_\pi^{\,2}}{24 F_\pi^{\,2}} (\boldsymbol{\phi} \cdot \boldsymbol{\phi})^2 + \ldots\end{aligned} \tag{A10}$$

The simplest form of the effective pion Lagrangian that has the proper chiral transformation properties is then

$$\mathscr{L}_{\text{eff}} = \frac{1}{4} F_\pi^{\,2} \operatorname{Tr} \partial_\mu U \partial^\mu U^\dagger + \frac{F_\pi^{\,2} m_\pi^{\,2}}{4} \operatorname{Tr}(U + U^\dagger - 2). \tag{A11}$$

In quadratic order, this reproduces the free pion Lagrangian

$$\mathscr{L}_{\text{eff}}^{(2)} = \frac{1}{2} \partial_\mu \boldsymbol{\phi} \cdot \partial^\mu \boldsymbol{\phi} - \frac{1}{2} m_\pi^{\,2} \boldsymbol{\phi} \cdot \boldsymbol{\phi}. \tag{A12}$$

However, the strictures of chiral invariance also imply dynamical content in higher orders. Thus, the quartic component of eq. (A11) is

$$\mathscr{L}_{\text{eff}}^{(4)} = \frac{1}{6F_\pi^2}\,\boldsymbol{\phi}\cdot\boldsymbol{\phi}\boldsymbol{\phi}\cdot\Box\boldsymbol{\phi} + \frac{1}{2F_\pi^2}(\partial_\mu\boldsymbol{\phi}\cdot\boldsymbol{\phi})^2 + \frac{m_\pi^2}{24F_\pi^2}\,\boldsymbol{\phi}\cdot\boldsymbol{\phi}\boldsymbol{\phi}\cdot\boldsymbol{\phi}, \quad \text{(A13)}$$

which implies the form

$$\begin{aligned} T(q_a,q_b;\,q_c,q_d) = {} & \frac{1}{F_\pi^2}(\delta^{ab}\delta^{cd}(-s+m_\pi^2) \\ & + \delta^{ac}\delta^{bc}(-t+m_\pi^2) + \delta^{ad}\delta^{bc}(-u+m_\pi^2)) \\ & - \frac{1}{3F_\pi^2}(4m_\pi^2 - q_a^2 - q_b^2 - q_c^2 - q_d^2) \\ & \times (\delta^{ab}\delta^{cd} + \delta^{ac}\delta^{bd} + \delta^{ad}\delta^{bc}) \end{aligned} \quad \text{(A14)}$$

for the $\pi\pi$ scattering amplitude. At threshhold eq. (A14) yields the famous Weinberg S-wave $\pi\pi$ scattering lengths [78],

$$a_0{}^0 = \frac{7m_\pi}{32\pi F_\pi^2} \qquad a_0{}^2 = -\frac{m_\pi}{16\pi F_\pi^2}, \quad \text{(A15)}$$

which are roughly borne out experimentally.

Obviously, higher-order chiral invariant combinations are also possible, e.g.,

$$\frac{\varepsilon^2}{4}\,\mathrm{Tr}[\partial_\mu U U^\dagger, \partial_\nu U U^\dagger]^2, \qquad \frac{\gamma^2}{4}\,[\mathrm{Tr}\;\partial_\mu U \partial^\mu U^\dagger]^2 \quad \text{(A16)}$$

A priori there is no condition that serves to normalize the coefficients of such terms. However, on theoretical grounds one anticipates that higher-order pieces are suppressed by factors [79]

$$\left(\frac{q^2}{\Lambda^2}\right),$$

where

$$\Lambda \sim 4\pi F_\pi \quad \text{(A17)}$$

is a scale parameter. This is also verified empirically in that fourth-order, i.e., four derivative, terms such as these generate d-wave pion scattering contributions. In fact, of the many possible forms for fourth-order terms, it can be shown that only two are linearly independent. Choosing these, as in eq. (A15), and fitting to observed d-wave scattering lengths [80]

$$a_2{}^0 = (17 \pm 3) \times 10^{-4} m_\pi^{-4}; \qquad a_2{}^2 = (1.3 \pm 3) \times 10^{-4} m_\pi^{-4} \tag{A18}$$

we find

$$\Lambda \sim 1.3 \text{ GeV}, \tag{A19}$$

in rough agreement with theoretical expectations.

Having seen how mesons can be accommodated within a chiral-invariant framework, we can ask how baryons are to be treated—where is the nucleon in such a formalism? The answer is somewhat surprising: it is already present in the effective meson Lagrangian in terms of quantum fluctuations about a "soliton" solution! A soliton is a solution of a nonlinear Lagrangian that is localized in space and propagates without dispersion—a lump of energy that moves through spacetime without changing its shape. Such soliton solutions are well known in fluid mechanics and in other nonlinear systems, and the possible relevance to nuclear physics was first suggested by Skyrme in 1960 [81]. In this regard, the quartic terms in the effective Lagrangian play a critical role. This is apparent since if *only* the simple quadratic piece of the Lagrangian were present, the energy of such a static solution,

$$E = \int d^3x \frac{F_\pi{}^2}{4} \operatorname{Tr} \nabla U \cdot \nabla U^\dagger, \tag{A20}$$

would be unstable under a scale transformation; that is, under the scale change $x \to \Lambda x$, we find

$$E \to \Lambda^{-1} E, \tag{A21}$$

i.e., it is energetically favorable to rescale the solution to infinite size, in which case the energy vanishes. Addition of the quartic terms stabilizes the soliton in that the energy equation then scales as

$$E \to \Lambda^{-1} E_{(2)} + \Lambda E_{(4)}, \tag{A22}$$

which can have a stable minimum when

$$E_{(4)} = E_{(2)}. \tag{A23}$$

An explicit form for the solution can be found by making the Skyrme ansatz

$$U(\mathbf{x}) = \exp(i\boldsymbol{\tau} \cdot \hat{x}F(x)), \tag{A24}$$

which couples spatial and isotopic degrees of freedom. Demanding that $U(\mathbf{x})$ obey the classical (Euler-Lagrange) equation of motion yields a differential equation for $F(x)$ (with $\gamma = 0$ as employed by Skyrme),

$$\begin{aligned} F'' + \frac{2}{x} F' - \frac{\sin 2F}{x^2} + 8 \frac{\varepsilon^2}{F_\pi{}^2} \bigg[& \frac{\sin 2F \sin^2 F}{x^4} \\ & - \frac{F'^2 \sin 2F}{x^2} - \frac{2F'' \sin^2 F}{x^2} \bigg] = 0, \end{aligned} \tag{A25}$$

which can be solved numerically, using the boundary conditions [82]

$$F(0) = \pi \qquad F(\infty) = 0. \tag{A26}$$

The quantum numbers associated with such a solution are indicated by the observation that

$$[\mathbf{K}, U(\mathbf{x})] = 0, \tag{A27}$$

where

$$\mathbf{K} = \mathbf{J} + \mathbf{I} \tag{A28}$$

and that $U(\mathbf{x})$ is invariant under the parity operation

$$U(\mathbf{x},t) \xrightarrow[P]{} U^\dagger(-\mathbf{x},t) = U(\mathbf{x},t). \tag{A29}$$

This suggests the soliton is a K-spin singlet and consists of a combination of states with $I = J$ and positive parity. It was conjectured by Skyrme and proven by Witten that the soliton changes its phase by π radians under a 2π rotation and is thus a fermion. The so-called Skyrmion represents then an admixture of nucleon ($I = J = \frac{1}{2}$) and delta ($I = J = \frac{3}{2}$) states. One quantizes the theory by adiabatically spinning the static solution in isospace via

$$U = A(t)U_0(\mathbf{x})A^{-1}(t), \tag{A30}$$

and then projecting out the component with definite total angular momentum. The effective Lagrangian becomes

$$L = -M + I\mathrm{Tr}\dot{A}\dot{A}^{-1} + \mathcal{O}(A^3), \tag{A31}$$

where in the simple Skyrme picture with $\gamma = 0$

$$I = \frac{8\pi}{3} F_\pi^2 \int_0^\infty dx\, x^2 \sin^2 F \left[1 + \frac{8\varepsilon^2}{F_\pi^2}\left(F'^2 + \frac{1}{x^2}\sin^2 F\right)\right] \tag{A32}$$

is the soliton "moment of inertia," and

$$\begin{aligned} M = 4\pi \int_0^\infty dx\, x^2 \Bigg\{ & \frac{F_\pi^2}{2}\left[F'^2 + \frac{2\sin^2 F}{x^2}\right] \\ & + 4\varepsilon^2 \frac{\sin^2 F}{x^2}\left[2F'^2 + \frac{\sin^2 F}{x^2}\right]\Bigg\} \end{aligned} \tag{A33}$$

is the mass associated with the classical solution. Using this action the nucleon and delta masses can be calculated, yielding results within 20% or so of their experimental values [83]!

Because of this success, much effort has been exerted in Skyrme model calculations [84]. Evaluations of nucleon static parameters and even of the nucleon-nucleon interaction have become commonplace. My own view in this regard is that the success of the Skyrme ansatz provides evidence for the importance and validity of chiral symmetry and in some sense for the validity of QCD itself. In view of the ad hoc nature of the solution and the semiclassical quantization of it, we should not take the specific numerical results too seriously. Nevertheless, the existence of this quantized soliton solution could conceivably help to explain one of the real mysteries of nuclear physics—where are the quarks? That is to say, in a conventional quark picture the nucleon is a state of three spin $\frac{1}{2}$ quarks bound into a roughly spherical volume of radius 0.8 fermis or so. Mesons, such as the pion or rho, are quark-antiquark states confined within a slightly smaller volume. In this sense there would exist tremendous overlap between two nucleons at a relative separation of, say, 0.7 fm or less (cf. Figure 4.6) It is then remarkable that a meson exchange picture is successful in predicting the basic nature of the nucleon-nucleon interaction. Of course, the pion exchange component has a range considerably larger than the nuclear radius and thus there is no mystery in this case. However, in the case of

$$\rho\colon r_\rho \sim \frac{1}{m_\rho} \sim 0.3 \text{ fm} \qquad \omega\colon r_\omega \sim \frac{1}{m_\omega} \sim 0.3 \text{ fm} \tag{A34}$$

exchange, the overlap would seem to play a crucial role.

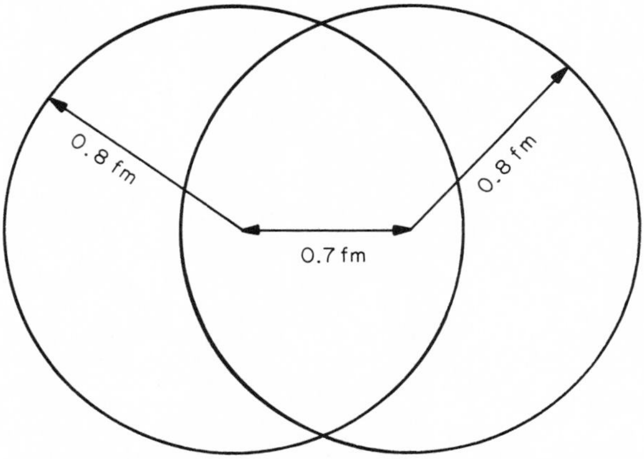

FIGURE 4.6. Sketched is the substantial overlap that arises when two spherical nucleons of radius 0.8 fm are stationed at a relative separation of 0.7 fm.

If, on the other hand, the effective chiral Lagrangian technique is used, the resulting picture is quite different. Quarks per se are not mentioned. The pseudoscalar fields are fundamental objects. The nucleon is a topological soliton, and the vector mesons can be considered as fundamental gauge fields in a locally chiral invariant model. The overlap problem is not really solved, but is less of a concern in such an approach. Of course, the real world is neither the simple quark model nor the picture offered by the chiral Lagrangian but is rather a fusion of the two approaches. Thus, both views are useful, though depending on the process being discussed, one or the other way may be favored.

Chiral symmetry has additional implications for nuclear physics which are at the present time puzzling. These have to do with the origin of the nucleon mass. Thus, writing

$$m_N = \langle N|\mathscr{H}_0 + m_u\bar{u}u + m_d\bar{d}d + m_s\bar{s}s|N\rangle, \tag{A35}$$

where

$$m_N^{(0)} \equiv \langle N|\mathscr{H}_0|N\rangle \tag{A36}$$

is the nucleon mass in the chiral symmetric limit ($m_u = m_d = m_s = 0$), we can divide the mass into three components

$$m_N = m_N^{(0)} + \delta m_s + \delta m_{ud}, \tag{A37}$$

with

$$\delta m_s = m_s \langle N|\bar{s}s|N\rangle \tag{A37}$$

and

$$\delta m_{ud} = \frac{1}{2}(m_u + m_d)\langle N|\bar{u}u + \bar{d}d|N\rangle \tag{A39}$$

being contributions to the nucleon mass arising from chiral $SU(3)_L \times SU(3)_R$ and $SU(2)_L \times SU(2)_R$ breaking corrections, respectively. Since a valence quark census of the nucleon reveals that only u,d quarks are present

$$|p\rangle \sim |uud\rangle \quad \text{and} \quad |n\rangle \sim |udd\rangle, \tag{A40}$$

one might suspect that $m_N^{(0)}$ is the dominant component of the nucleon mass, with the chiral breaking term δm_{ud} playing a significant but smaller role, and δm_s being essentially negligible. These expectations are not borne out, however, as I shall demonstrate.

It turns out that $\delta m_{u,d}$ can be measured, albeit somewhat indirectly, in pion-nucleon scattering. The basic idea is that via the PCAC condition, eq. (4.123), the πN scattering amplitude can be related in the unphysical limit $q_1,q_2 \to 0$ to the matrix element of a commutator

$$\begin{aligned} \langle \pi^a_{q_2} N_{p_2}(\text{out})|\pi^b_{q_1} N_{p_1}(\text{in})\rangle &\xrightarrow[q_1,q_2 \to 0]{} \frac{1}{F_\pi^{\;2}} \\ &\times \left\langle N_{p_2}\left|\left[\int d^3x A_0{}^a(\mathbf{x},t),\, \partial^\mu A_\mu{}^b(0)\right]\right|N_{p_1}\right\rangle, \end{aligned} \tag{A41}$$

which, in turn, using the quark equations of motion

$$\begin{aligned} i\partial^\mu A_\mu{}^b(x) &= i\partial^\mu \bar{\psi}(x)\gamma_\mu\gamma_5 \frac{1}{2}\tau^b\psi(x) \\ &= \bar{\psi}(x)i\overleftarrow{\not\partial}\gamma_5 \frac{1}{2}\tau^b\psi(x) + \bar{\psi}(x)i\overrightarrow{\not\partial}\gamma_5 \frac{1}{2}\tau^b\psi(x) \\ &= -\bar{\psi}(x)\gamma_5\left\{M, \frac{1}{2}\tau^b\right\}\psi(x), \end{aligned} \tag{A42}$$

where

$$M = \begin{pmatrix} m_u & 0 \\ 0 & m_d \end{pmatrix} \tag{A43}$$

is the quark mass matrix, can be related to δm_{ud} since

$$i\left[\int d^3x A_0{}^a(\mathbf{x}, t), \partial^\mu A_\mu{}^b(0)\right] = \frac{1}{2}(m_u + m_d)(\bar{u}u + \bar{d}d)\delta^{ab}. \tag{A44}$$

We can extrapolate to the physical pion limit by use of chiral symmetry and dispersion relations, yielding [85]

$$\delta m_{ud} = 55 \pm 7 \text{ MeV}. \tag{A45}$$

Thus, we find that this component of the nucleon mass is very small.

The $SU(3)_L \otimes SU(3)_R$ breaking parameter δm_s can be evaluated via a somewhat different route. Since the masses of the $J^p = \frac{1}{2}^+$ baryons are given by

$$m_B = \langle B|m_u\bar{u}u + m_d\bar{d}d + m_s\bar{s}s|B\rangle + m_N^{(0)}, \tag{A46}$$

it is possible in the limit of SU(3) symmetry, assuming the u,d,s quark masses are small, to relate them, yielding [86]

$$\begin{aligned}\Delta \equiv \frac{1}{2}(m_u + m_d)\langle N|\bar{u}u + \bar{d}d - 2\bar{s}s|N\rangle &= \frac{3(m_u + m_d)}{m_u + m_d - 2m_s}(m_\Lambda - m_\Xi) \\ &\cong 25 \text{ MeV}.\end{aligned} \tag{A47}$$

Comparison with eqs. (A39), (A45) indicates that there exists an inconsistency unless the nucleon has a substantial $\bar{s}s$ matrix element

$$\begin{aligned}\delta m_s = m_s\langle N|\bar{s}s|N\rangle &= -\frac{m_s}{m_u + m_d}(\Delta - \delta m_{ud}) \\ &= 375 \pm 90 \text{ MeV}.\end{aligned} \tag{A48}$$

By subtraction we then obtain

$$m_N^{(0)} = 510 \pm 90 \text{ MeV} \tag{A49}$$

as the chiral symmetric component. This result appears to violate one's intuition about the importance of the strange quark to the structure of the nucleon. Nevertheless, it is an inescapable consequence of the chain of steps given above.

A number of physicists have worried about the significance of this result and have attempted to find an "escape route" [87]. One possibility is that while chiral $SU(2)_L \otimes SU(2)_R$ is a nearly good symmetry with deviations due to m_u,m_d able to be treated perturbatively, the same may not be true of chiral $SU(3)_L \otimes SU(3)_R$. Indeed the relatively large mass of the s-quark—~ 150 MeV—gives one pause about the use of $m_s\bar{s}s$ in first-order perturbation theory. However, splittings of the $J^P = \frac{1}{2}^+$ baryon masses due to this term are relatively small:

$$\frac{m_\Lambda - m_p}{m_\Lambda + m_p} \sim 10\%; \tag{A50}$$

and use of simple perturbation theory yields the Gell-Mann–Okubo mass formula which successfully relates the baryon masses [88]

$$\begin{aligned} m_N + m_\Xi &= \frac{1}{2}(m_\Lambda + 3m_\Sigma) \\ 2260\ \text{MeV} \quad & \quad 2340\ \text{MeV}. \end{aligned} \tag{A51}$$

A second possible resolution of this problem is the inclusion of so-called chiral loop corrections that account for the effects of higher-order perturbative corrections to the effective chiral Lagrangian. Estimates indicate that in this way $m_N^{(0)}$ is raised by about 100 MeV and δm_s is lowered by about the same amount [89]. However, a significant strange quark component is still required.

Actually, at a qualitative level, the presence of a sizable s-quark effect should not really be a surprise, since the existence of a kaon cloud around the nuclear core should certainly arise from the virtual processes

$$\begin{aligned} p &\rightarrow \Sigma^+ K^0, \Sigma^0 K^+, \Lambda K^+ \\ n &\rightarrow \Sigma^- K^+, \Sigma^0 K^0, \Lambda K^0. \end{aligned}$$

Since Σ,Λ,K states do contain strange quarks/antiquarks, the existence of a significant $\bar{s}s$ matrix element for the nucleon is a natural consequence. However, careful numerical evaluations of the expected size of the effect have not yet been performed.

It is possible, of course, to test this assertion directly by means of experiment, although this will not be a simple task. For example, one could in principle use K-nucleon scattering as a probe for $\langle N|\bar{s}s|N\rangle \neq 0$ [90].

In this case the unphysical unit $q_1,q_2 \to 0$ yields

$$\langle K^a_{q_2} N_{p_2}(\text{out}) | K_q{}^b N_{p_1}(\text{in}) \rangle \xrightarrow[q_1,q_2 \to 0]{} -\frac{1}{F_K{}^2} \times \left\langle N_{p_2} \left| \left[\int d^3x A_0{}^a(\mathbf{x},t), \partial^\mu A^b_\mu(0) \right] \right| N_{p_1} \right\rangle, \tag{A52}$$

and since

$$\begin{aligned} i\left[\int d^3x A_0^{K^-}(\mathbf{x},t), \partial^\mu A_\mu^{K^+}(0) \right] &= \frac{1}{2}(m_u + m_s)(\bar{u}u + \bar{s}s) \\ i\left[\int d^3x A_0^{K^0}(\mathbf{x},t), \partial^\mu A_\mu^{K^0}(0) \right] &= \frac{1}{2}(m_d + m_s)(\bar{d}d + \bar{s}s) \end{aligned} \tag{A53}$$

offer an experimental window to the $\langle N|\bar{s}s|N\rangle$ matrix element. The problem here is that the required extrapolation from the physical region of KN scattering is a long one and consequently of dubious reliability.

A second possible experiment to probe for strange quark content of the nucleon involves the weak interaction. The point is that the weak neutral current has the form

$$\begin{aligned} J_\mu = \bar{u}_L\gamma_\mu u_L - \bar{d}_L\gamma_\mu d_L + \bar{c}_L\gamma_\mu c_L - \bar{s}_L\gamma_\mu s_L + \dots \\ - \sin^2\theta_w\left(\frac{2}{3}\bar{u}\gamma_\mu u - \frac{1}{3}\bar{d}\gamma_\mu d + \frac{2}{3}\bar{c}\gamma_\mu c - \frac{1}{3}\bar{s}\gamma_\mu s + \dots\right). \end{aligned} \tag{A54}$$

Since the light quark combination $\bar{u}u - \bar{d}d$ is an isovector, and since the charmed quark is much too heavy to play a significant nuclear role, we observe that the only source of an isoscalar axial vector current is the term $\bar{s}\gamma_\mu\gamma_5 s$. Thus to the extent that one could detect such an effect, the experiment would offer an independent strange quark measurement.

A possible idea would be to utilize an electron accelerator with a polarized e^- beam and to look for a difference in the cross section of electrons polarized parallel vs. antiparallel with respect to the beam direction when they scatter from an isoscalar target, e.g., ^{12}C. As I will discuss in more detail in the next chapter, such a difference can arise only from violation of parity invariance, which in this case comes from interference between

hadron		*electron*
isoscalar axial current	and	polar vector current
	or	
isoscalar polar vector current	and	axial vector current.

Thus sensitivity to the isoscalar axial does exist. However, in the Weinberg-Salam model the electron polar vector coupling is proportional to [91]

$$1 - 4\sin^2\theta_w \ll 1,$$

making this probe of strange quark content virtually useless. A way around this dilemma is provided by use of elastic neutrino scattering

$$\nu_e + {}^{12}\mathrm{C} \rightarrow \nu_e + {}^{12}\mathrm{C}.$$

However, limits in beam intensity and uncertainties in flux make this approach also rather challenging. Nevertheless, recent elastic neutrino scattering measurements from protons and deuterium carried out at Brookhaven National Laboratory indicate the presence of an isoscalar axial component at about the 10% level [92], in agreement with earlier theoretical estimates [93].

I conclude that chiral symmetry has important implications for nuclear physics that are only beginning to be explored. The combination of Skyrme and quark model approaches and the analysis of chiral implication for scattering processes will doubtless provide trenchant insights, enabling a depth of understanding unachievable until now.

References

[1] R. P. Feynman and M. Gell-Mann, Phys. Rev. *109*, 193 (1958).

[2] R. P. Feynman, "*Surely You're Joking Mr. Feynman*!" *Adventures of a Curious Character*. Norton, New York (1985).

[3] See, e.g., R. E. Marshak, Riazuddin, and C. P. Ryan, *Theory of Weak Interactions in Particle Physics*, Wiley, New York (1969).

[4] B. R. Holstein, Rev. Mod. Phys. *46*, 789 (1974); B. R. Holstein and S. B. Treiman, Phys. Rev. *C3*, 1921 (1971).

[5] H. Behrens and J. Janecke, *Numerical Tables for Beta Decay and Electron Capture*, Landolt-Bornstein, New Series, vol. I/4, Springer, Berlin (1969).

[6] D. H. Wilkinson and R. E. Marrs, Nucl. Inst. Meth. *105*, 505 (1972).

[7] A. Sirlin and R. Zucchini, Phys. Rev. Lett. *57*, 1994 (1986); W. Jaus and G. Rasche, Phys. Rev. *D35*, 3420 (1987); A. Sirlin, Phys. Rev. *D35*, 3423 (1987).

[8] A. Sirlin, Nucl. Phys. *B71*, 29 (1974), and Rev. Mod. Phys. *50*, 573 (1978).

[9] E. S. Abers, D. Dicus, R. Norton, and H. Quinn, Phys. Rev. *167*, 1461 (1968).

[10] W. J. Marciano and A. Sirlin, Phys. Rev. Lett. *56*, 22 (1986).

[11] H. Leutwyler and M. Roos, Z. Phys. *C25*, 91 (1984); J. F. Donoghue, B. R. Holstein, and S. Klimt, Phys. Rev. *D35*, 934 (1987).

[12] M. Gell-Mann, Phys. Rev. *111*, 362 (1988).

[13] F. P. Calaprice and B. R. Holstein, Nucl. Phys. *A273*, 301 (1976).

[14] Y. K. Lee, L. W. Mo, and C. S. Wu, Phys. Rev. Letters *10*, 253 (1963).

[15] F. P. Calaprice and B. R. Holstein, Nucl. Phys. *A273*, 301 (1976).

[16] C. W. Wu, Y. K. Lee, and L. W. Mo, Phys. Rev. Lett. *39*, 72 (1977).

[17] W. Kaina et al., Phys. Lett. *B70*, 411 (1977).

[18] F. P. Calaprice and D. E. Alburger, Phys. Rev. *C17*, 730 (1978).

[19] H. Genz et al., Nucl. Phys. *A267*, 13 (1976).

[20] L. van Elmbt, J. Deutsch, and R. Prieels, Nucl. Phys. *A469*, 531(1987).

[21] L. Szybisz and V. M. Silbergleit, J. Phys. *G7*, 1201 (1981).

[22] B. R. Holstein, Phys. Rev. *C29*, 623 (1984).

[23] B. R. Holstein and S. B. Treiman, Phys. Rev. *D13*, 3059 (1976).

[24] S. Weinberg, Phys. Rev. *112*, 1375 (1958).

[25] B. R. Holstein and S. B. Treiman, ref. 4; S. P. Rosen, Phys. Rev. *D5*, 760 (1972).

[26] D. H. Wilkinson and D. E. Alburger, Phys. Rev. Lett. *24*, 1134 (1970), and *26*, 2227 (1971), and Phys. Lett. *32B*, 190 (1970).

[27] D. H. Wilkinson, Phys. Lett. *B31*, 447 (1970).

[28] I. S. Towner, Nucl. Phys. *A216*, 589 (1973); D. H. Wilkinson, Phys. Rev. Lett. *27*, 1018 (1971).

[29] C. W. Kim, Phys. Lett, *B24*, 383 (1971); B. R. Holstein and S. B. Treiman, ref. 4. L. Grenacs, Ann. Rev. Nuccl. Sci. *35*, 455 (1985).

[30] R. E. Tribble and G. T. Garvey, Phys. Rev. *C12*, 967 (1975); R. D. McKeown et al., Phys. Rev. *C22*, 738 (1980); A. M. Nathan et al., Phys. Rev. Lett. *35*, 1137 (1975), and erratum *49*, 1056 (1982).

[31] H. Brandle et al., Phys. Rev. Lett. *40*, 306 (1978); P. Lebrun et al., Phys. Rev. Lett. *40*, 302 (1978); H. Brandle et al., Phys. Rev. Lett. *41*, 299 (1978); Y. Masuda et al., Phys. Rev. Lett. *43*, 1083 (1979). T. Minamisono et al., J. Phys. Soc. Jpn. Suppl. *55*, 1012 (1987).

[32] R. E. Tribble and D. P. May, Phys. Rev. *C18*, 2704 (1978); N. Rolin et al., Phys. Lett. *70B*, 23 (1977), and *79B*, 359 (1978); R. E. Tribble, D. P. May, and D. M. Towner, Phys. Rev. *C23*, 2245 (1981). R. D. Rosa et al., Phys. Rev. *C37*, 2722 (1988).

[33] F. P. Calaprice et al., Phys. Rev. Lett. *35*, 1566 (1975).

[34] D. Schreiber, Princeton University Ph.D. dissertation (1982).

[35] J. F. Donoghue and B. R. Holstein, Phys. Rev. *D25*, 206 (1982).

[36] K. Kubodera, J. Delorme, and M. Rho, Phys. Rev. Lett. *40*, 755 (1978).

[37] M.A.B. Beg, R. Budny, R. Mohapatra, and A. Sirlin, Phys. Rev. Lett. *38*, 1252 (1977); T. Rizzo and G. Senjanovic, Phys. Rev. Lett. *46*, 1315 (1981).
[38] B. R. Holstein and S. B. Treiman, Phys. Rev. *D16*, 2369 (1977).
[39] H. Paul, Nucl. Phys. *A154*, 160 (1970).
[40] J. Van Klinken et al., Phys. Rev. Lett. *50*, 94 (1983); V. A. Wichers et al., Phys. Rev. Lett. *58*, 1821 (1987).
[41] G. Gerber et al., Phys. Rev. *D15*, 1189 (1977).
[42] M. Skalsey et al., Phys. Rev. Lett. *49*, 708 (1982) and Phys. Rev. *C39*, 986 (1989), A. S. Carnoy et al., in *Weak and Electromagnetic Interactions in Nuclei*, ed. H. Klapdor, Springer-Verlag, New York (1987), p. 534; J. Deutsch, Proc. Workshop on the Breaking of Fundamental Symmetries in Nuclei, Santa Fi (1988).
[43] B. R. Holstein and S. B. Treiman, Phys. Rev. *D16*, 2364 (1977).
[44] F. P. Calaprice, in *Hyperfine Interactions IV*, North-Holland, Amsterdam (1978), p. 25.
[45] P. Bopp et al., J. de Physique Colloq. *C3*, 21 (1984), and in *Capture Gamma Ray Spectroscopy and Related Topics–1984*, ed. S. Raman, AIP Conf. Proc. No. 125, AIP, New York (1985), p. 881.
[46] A. S. Carnoy, J. Deutsch and B. R. Holstein, Phys. Rev. *D38*, 1636 (1988).
[47] M. A. Beg et al., ref. 37.
[48] J. Case et al., Phys. Rev. Lett. *51*, 627 (1983).
[49] D. P. Stoker et al., Phys. Rev. Lett. *54*, 1887 (1985); A. Jodidio et al, Phys. Rev. *D34*, 1967 (1986) and *D37*, 237 (E) (1988).
[50] I. Beltrami et al., Phys. Lett. *B194*, 326 (1987).
[51] G. Senjanovic and R. N. Mohapatra, Phys. Rev. *D12*, 1502 (1975).
[52] L. Wolfenstein, Phys. Rev. *D29*, 2130 (1984).
[53] Y. Nambu and Y. Hara, Phys. Rev. Lett. *16*, 874 (1966); C. Bouchiat and Ph. Meyer, Phys. Lett. *25B*, 282 (1967).
[54] J. F. Donoghue and B. R. Holstein, Phys. Lett. *113B*, 382 (1982).
[55] J. F. Donoghue, E. Golowich, B. R. Holstein, and W. Ponce, Phys. Rev. *D21*, 186 (1980).
[56] G. Beall, A. Soni, and M. Bander, Phys. Rev. Lett. *48*, 848 (1982).
[57] See, e.g., M. Weinberg, *Chiral Symmetry*, Springer Tracts in Modern Physics, vol. 60, Springer-Verlag, New York (1971).
[58] J. Goldstone, Nuovo Cimento *19*, 154 (1961); J. Goldstone, A. Salam, and S. Weinberg, Phys. Rev. *127*, 965 (1961).
[59] M. L. Goldberger and S. B. Treiman, Phys. Rev. *111*, 354 (1958).
[60] M. Gell-Mann and M. Levy, Nuovo Cimento *17*, 705 (1960); Y. Nambu, Phys. Rev. Lett. *4*, 380 (1960).
[61] B. R. Holstein, ref. 4.

[62] H. Primakoff, in *Muon Physics*, ed. by V. W. Hughes and C. S. Wu, Academic Press, New York (1975), vol. II.

[63] B. R. Holstein, ref. 4; H. Primakoff, ref. 62.

[64] Although this relation is not rigorously justified, it is approximately valid because both form factors scale with the nuclear radius. Also, the induced tensor plays only a small part in the capture rate, so any correction to eq. (4.111) should be numerically insignificant.

[65] L. B. Auerbach et al., Phys. Rev. *138*, B127 (1965); D. R. Clay et al., Phys. Rev. *140*, B586 (1965); C. W. Kim and H. Primakoff in *Mesons in Nuclei*, ed. by M. Rho and D. H. Wilkinson, North-Holland, Amsterdam (1979), vol. I.

[66] G. Bardin et al., Phys. Lett. *B104*, 320 (1981).

[67] G. H. Miller et al., Phys. Lett. *B41*, 50 (1972).

[68] A. Passoz et al., Phys. Lett. *B70*, 265 (1977).

[69] V. Roesch et al., Phys. Rev. Lett. *46*, 1507 (1981).

[70] B. R. Holstein, Phys. Rev. *C4*, 764 (1971).

[71] R. D. Hart et al., Phys. Rev. Lett. *39*, 399 (1977).

[72] A. Frischknecht et al., Phys. Rev. *C32*, 1506 (1885).

[73] M. Dobeli et al., Phys. Rev. *C37*, 1633 (1988).

[74] J. Delorme et al., Ann. Phys. (N.Y.) *102*, 273 (1976).

[75] See, e.g., R. J. Blin-Stoyle in *Mesons in Nuclei*, ed. M. Rho and D. H. Wilkinson, North-Holland, Amsterdam (1979), vol. I and references therein.

[76] B. A. Brown and B. H. Wildenthal, At. Data and Nucl. Data Tables *33*, 347 (1985); see also D. H. Wilkinson, Phys. Rev. *C7*, 930 (1973). Nucl. Phys. *A209*, 470 (1973), and Nucl. Phys. *A225*, 315 (1974).

[77] See, e.g., S. Gasiorowicz and D. A. Geffen, Rev. Mod. Phys. *41*, 531 (1969).

[78] S. Weinberg, Phys. Rev. Lett. *16*, 879 (1966), and *17*, 336 (1966).

[79] A. Manohar and H. Georgi, Nucl. Phys. *B234*, 189 (1984); J. F. Donoghue, E. Golowich, and B. R. Holstein, Phys. Rev. *D30*, 587 (1984).

[80] W. Hoogland et al., Nucl. Phys. *B69*, 266 (1974); B. Hyams et al., Nucl. Phys. *B77*, 202 (1974); J. Baton et al., Phys. Lett. *B33*, 528 (1970); N. B. Durosy et al., Phys. Lett. *B45*, 517 (1973); N. Biswas et al., Phys. Rev. Lett. *47*, 1378 (1981); N. Cason et al., Phys. Rev. Lett. *48*, 1316 (1982); J. F. Donoghue, E. Golowich, and B. R. Holstein, Phys. Rev. Lett. *53*, 747 (1984).

[81] T.H.R. Skyrme, Proc. Roy. Soc. *A260*, 127 (1961).

[82] The boundary condition at $r = 0$ is that required for the solution to have unit baryon number.

[83] G. Adkins. C. Nappi, and E. Witten, Nucl. Phys. *B228*, 552 (1983); I. Zahed and G. E. Brown, Phys. Rept. *142*, 1 (1986).

[84] See, e.g., *Workshop on Nuclear Chromodynamics*, ed. S. Brodksy and E. Moniz, World Scientific, Singapore (1986), Ch. 5.

[85] R. L. Jaffe, Phys. Rev. *D21*, 3215 (1980); H. J. Pagels and W. J. Pardee, Phys. Rev. *D4*, 3335 (1971); P. M. Gensini, Nuovo Cimento *60A*, 221 (1980).

[86] J. F. Donoghue and C. R. Nappi, Phys. Lett. *B168*, 105 (1986); T. P. Cheng, Phys. Rev. *D13*, 2161 (1976).

[87] R. L. Jaffe, Nucl. Phys. *A478*, 3c (1988).

[88] S. Okubo, Prog. Theo. Phys. 949 (1962); M. Gell-Mann, R. J. Oakes, and B. Renner, Phys. Rev. *175*, 2195 (1968).

[89] J. Gasser, Nucl. Phys. *B279*, 65 (1987); E. Reya, Rev. Mod. Phys. *46*, 545 (1974).

[90] R. L. Jaffe, Comm. Nucl. Part. Phys. *17*, 163 (1987).

[91] See, however, R. D. McKeown, Phys. Lett. *B219*, 140 (1989).

[92] K. Abe et al., Phys. Rev. Lett. *56*, 1107, 1183(E) (1986); L. A. Ahrens et al., Phys. Rev. *D35*, 785 (1987).

[93] J. E. Kim et al., Rev. Mod. Phys. *53*, 211 (1981); L. Wolfenstein, Phys. Rev. *D19*, 3450 (1979); J. Collins et al., Phys. Rev. *D18*, 242 (1978).

Chapter 5

P, T, AND CP

Backward, turn backward, O time in your flight,
Make me a child again just for tonight.
—ELIZABETH AKERS ALLEN

"I am right and you are right
and all is right as right can be."
—W. S. GILBERT

5.1 Time-Reversal Invariance

One of the most important results in physics is that nearly all processes obey the principle of time reversal, or T-invariance. Perhaps the easiest way to visualize this concept is to imagine that a film of some physical process is being screened before a skeptical audience of physicists—say, a simple scattering reaction as illustrated in Figure 5.1. The assembled scientists are allowed to analyze the reaction with meter sticks, stop-watches, etc., and will presumably affirm that the film is consistent with the laws of physics. Now, however, imagine that (unknown to the audience) the projectionist runs the film backwards. If the assembled physicists cannot detect any violation of physical law, i.e., cannot *prove* from their observations that the film is being shown in reverse, then the system is said to be T-invariant. In the case of the simple scattering process shown in Figure 5.1, the time reversal symmetry seems clear. However, imagine instead a motion picture of a glass falling to the floor and shattering. Am I implying that the assembled physicists would not be able to tell whether the film was shown backwards by the projectionist? Surprisingly yes! The point is that *if* the initial conditions were right—*all* shards of glass moving upward with just correct velocities—the glass would indeed reassemble. There is no violation of the laws of physics. Of course, having these very precise initial conditions is *extremely* unlikely (but not impossible), and that is associated with the second law of thermodynamics. Thus the physicists would presume that the film of the glass reassembling was running backward. However, they could not find "smoking gun" evidence in the form of violation of some fundamental (nonstatistical) physical law.

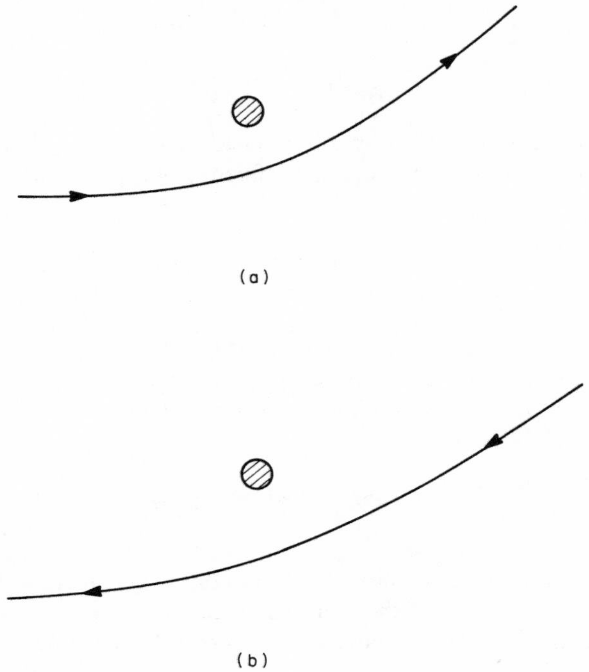

FIGURE 5.1. A simple scattering process (a) and its time reversed analog (b).

Classically, the time reversal operation corresponds to the replacements

$$t \to -t$$
$$\mathbf{r} \to \mathbf{r}$$

and hence

$$\mathbf{v} = \frac{d\mathbf{r}}{dt} \to \frac{d\mathbf{r}}{d(-t)} = -\mathbf{v} \tag{5.1}$$
$$\mathbf{a} = \frac{d^2\mathbf{r}}{dt^2} \to \frac{d^2\mathbf{r}}{d(-t)^2} = \mathbf{a}$$
$$\mathbf{L} = \mathbf{r} \times m\mathbf{v} \to \mathbf{r} \times (-m\mathbf{v}) = -\mathbf{L}.$$

So that the total angular momentum

$$\mathbf{J} = \mathbf{L} + \mathbf{S} \tag{5.2}$$

behaves consistently under T, we must also require

$$\mathbf{S} \to -\mathbf{S}. \tag{5.3}$$

Under this set of operations, it is clear that classical physics is invariant. Thus, for motion under the influence of a potential, we have

$$m\frac{d^2\mathbf{r}}{dt^2} = -\nabla V(r) \to m\frac{d^2\mathbf{r}}{d(-t)^2} = -\nabla V(r). \tag{5.4}$$

That one must exercise caution in application of time reversal can be seen by considering motion as analyzed by an observer on the surface of the earth, who employs the equation of motion [1]

$$m\frac{d^2\mathbf{r}}{dt^2} = \mathbf{F}(\mathbf{r}) + m\mathbf{g} - 2m\boldsymbol{\omega} \times \mathbf{v} - m\boldsymbol{\omega} \times (\boldsymbol{\omega} \times \mathbf{r}), \tag{5.5}$$

where $\boldsymbol{\omega}$ is the angular velocity of rotation of the earth. Here the velocity-dependent contribution is the Coriolis term, while the piece quadratic in $\boldsymbol{\omega}$ is the centrifugal force. Under time reversal this becomes

$$m\frac{d^2\mathbf{r}}{dt^2} = \mathbf{F}(\mathbf{r}) + m\mathbf{g} + 2m\boldsymbol{\omega}_T \times \mathbf{v} - m\boldsymbol{\omega}_T \times (\boldsymbol{\omega}_T \times \mathbf{r}), \tag{5.6}$$

which seemingly is T-noninvariant. However, recall that under time reversal *all* velocities are reversed, so that

$$\omega \to \omega_T = -\omega, \tag{5.7}$$

i.e., the earth itself must reverse its rotation, and time reversal invariance is restored. A related situation arises with respect to the Lorentz force law,

$$m\frac{d^2\mathbf{r}}{dt^2} = e(\mathbf{E} + \mathbf{v} \times \mathbf{B}) \to m\frac{d^2\mathbf{r}}{d(-t)^2} = e(\mathbf{E}_T - \mathbf{v} \times \mathbf{B}_T). \tag{5.8}$$

Assuming $\mathbf{E}_T = \mathbf{E}$, $\mathbf{B}_T = \mathbf{B}$ produces a noninvariant equation. However, recall that a magnetic field is produced by the motion of electromagnetic currents, which are themselves reversed under $t \to -t$. Thus we require $\mathbf{B}_T = -\mathbf{B}$ and we see that the Lorentz force law is indeed time reversal symmetric.

Similarly, quantum mechanics can be shown to be T-invariant (provided the interaction potentials are real). For a spin-$\frac{1}{2}$ particle interacting with an external potential and a spin-orbit force, the Schrödinger equation is

$$\begin{gathered} i\frac{\partial\psi}{\partial t} = \left(-\frac{1}{2m}\nabla^2 + V_1(r) + \boldsymbol{\sigma}\cdot\mathbf{L}V_2(r)\right)\psi \\ \Big\downarrow{\scriptstyle t\to -t} \\ -i\frac{\partial\psi}{\partial t} = \left(-\frac{1}{2m}\nabla^2 + V_1(r) + \boldsymbol{\sigma}\cdot\mathbf{L}V_2(r)\right)\psi, \end{gathered} \tag{5.9}$$

which appears to be noninvariant. Now, however, define the time-reversed wave function via

$$\psi_T(\mathbf{x},t) = U\psi^*(\mathbf{x},-t), \tag{5.10}$$

where

$$U = i\sigma_2 = \begin{pmatrix} 0 & 1 \\ -1 & 0 \end{pmatrix} \tag{5.11}$$

is a unitary transformation. Then the time-reversed equation becomes

$$i\frac{\partial\psi_T}{\partial t} = \left(-\frac{1}{2m}\nabla^2 + V_1(r) + \boldsymbol{\sigma}\cdot\mathbf{L}V_2(r)\right)\psi_T \tag{5.12}$$

and *is* evidently T-invariant. That the time reversed wave function should have this form is suggested by beginning with a plane wave eigenstate that has its polarization along the positive z-direction,

$$\psi(x,t) = e^{i\mathbf{p}\cdot\mathbf{x}-iEt}\begin{pmatrix} 1 \\ 0 \end{pmatrix}. \tag{5.13}$$

The time-reversed wave function is then

$$\psi_T(\mathbf{x},t) = i\sigma_2\psi^*(\mathbf{x},-t) = -e^{-i\mathbf{p}\cdot\mathbf{x}-iEt}\begin{pmatrix} 0 \\ 1 \end{pmatrix}, \tag{5.14}$$

which has reversed momentum and spin, as expected classically. (Note that this combination of complex conjugation together with a unitary operator is both nonlinear *and* nonunitary. Wigner has referred to it as an "anti-unitary" transformation [2].)

What does theory have to say about the time reversal properties of more general interactions? As discussed earlier, the "standard" six-quark model *does* have room for T-violation—if $\delta \neq 0, \pi$—since the KM matrix becomes complex. However, this T-violating phase is only associated with the heavy quark sector, and no T-violation is expected to be seen in low energy nuclear interactions. This is only one such model of T-violation—usually called the KM model, for obvious reasons. However, it is particularly important, since it is the *only* model of T-violation that arises naturally within the context of already known physics.

It is also interesting to consider other possibilities, since we should be aware of the possible alternatives to the KM picture and the way in which experiment might distinguish between models. This exercise is more than academic, because of the famous CPT theorem of Luders and Pauli, which states that under the following conditions, the product of charge conjugation with parity with time reversal *must* be a good symmetry [3]:

1. Lorentz invariance (all inertial observers should see the same physics).
2. Locality. (For example, when charge disappears here and appears there, it has to pass through the space in between. This is achieved by the existence of a *locally* conserved current j_μ with

$$\partial^\mu j_\mu = 0. \tag{5.15}$$

Integration over a finite volume and use of Gauss's theorem gives

$$\frac{d}{dt}\int_V j_0 d^3x = -\int_V \nabla \cdot \mathbf{j} d^3x = -\int_A \mathbf{j} \cdot d\mathbf{S}, \tag{5.16}$$

which says that any charge that disappears from the inside must have flowed through the surface.)
3. Spin-statistics (integer spin fields commute and obey Bose statistics, while half integer spin fields anti-commute and obey Fermi statistics).

In 1964, Christenson, Cronin, Fitch, and Turlay observed the decay of the long-lived kaon, K_L, into the same two-pion final state as does the short-lived kaon, K_S, which demonstrated that CP was violated [4]. To understand why this is the case, note that

$$\mathrm{CP}|K^0\rangle = |\bar{K}^0\rangle, \qquad \mathrm{CP}|\bar{K}^0\rangle = |K^0\rangle \tag{5.17}$$

where we have used the freedom to make a strangeness gauge transformation

$$|K^0\rangle \to e^{i\alpha S}|K^0\rangle = e^{i\alpha}|K^0\rangle$$
$$|\bar{K}^0\rangle \to e^{i\alpha S}|\bar{K}^0\rangle = e^{-i\alpha}|\bar{K}^0\rangle \tag{5.18}$$

to define the phases of these states, so that

$$C|K^0\rangle = -|\bar{K}^0\rangle$$
$$C|\bar{K}^0\rangle = -|K^0\rangle. \tag{5.19}$$

(Introduction of an arbitrary phase between states of different strangeness is always possible since corresponding amplitudes cannot be made to interfere.) Also, under $C \otimes P$

$$\mathrm{CP}|\pi^0\pi^0\rangle = |\pi^0\pi^0\rangle; \qquad \mathrm{CP}|\pi^0\pi^0\pi^0\rangle = -|\pi^0\pi^0\pi^0\rangle. \tag{5.20}$$

Hence, if the weak interaction H_w is CP-invariant, only the linear combination

$$|K_1{}^0\rangle = \frac{1}{\sqrt{2}}(|K^0\rangle + |\bar{K}^0\rangle), \tag{5.21}$$

which satisfies

$$\mathrm{CP}|K_1{}^0\rangle = |K_1{}^0\rangle, \tag{5.22}$$

can decay via H_w to $|\pi^0\pi^0\rangle$, while the linear combination

$$|K_2{}^0\rangle = \frac{1}{\sqrt{2}}(|K^0\rangle - |\bar{K}^0\rangle) \tag{5.23}$$

can decay only to $|\pi^0,\pi^0,\pi^0\rangle$. Since there is more phase space available for $|\pi^0\pi^0\rangle$, the 2π decay will proceed at a much faster rate than its 3π counterpart. Thus the short-lived kaon K_S having $\tau_S = 9 \times 10^{-11}$ sec is identified with $|K_1{}^0\rangle$, while the K_L having $\tau_L = 5 \times 10^{-8}$ sec is identified with $|K_2{}^0\rangle$. Only if CP is violated—i.e., if the physical K_S, K_L states are mixtures of CP eigenstates $K_1{}^0$, $K_2{}^0$

$$|K_L\rangle = |K_2{}^0\rangle + \varepsilon|K_1{}^0\rangle$$
$$|K_S\rangle = |K_1{}^0\rangle + \varepsilon|K_2{}^0\rangle, \tag{5.24a}$$

or if a direct weak CP-violating transition

$$\langle \pi^0\pi^0 | H_W | K_2{}^0 \rangle \neq 0 \tag{5.24b}$$

is possible can *both* K_L and K_S decay to a two-pion state as seen in the Fitch-Cronin experiment. The signal was actually in the $\pi^+\pi^-$ channel, for which similar though not identical arguments obtain, and subsequent measurements involving $\pi^0\pi^0$ decay showed that the transition occurs predominantly through the mixing scenario. The observed mixing is very small, i.e.,

$$\varepsilon \sim 2 \times 10^{-3}, \tag{5.25}$$

so the strength of the CP violating interaction is extraordinarily weak:

$$\frac{H(\text{CP-violating})}{H_{\text{strong}}} \sim Gm_p{}^2 \times 10^{-3} \sim 10^{-8}. \tag{5.26}$$

Nevertheless, its presence is definitely detected, and therefore by the CPT theorem, there must exist a corresponding violation of time reversal invariance. At the present time this T-violation has been detected within the $K^0 - \bar{K}^0$ system and nowhere else.

Recent experiments comparing $K_L \to \pi^0\pi^0$ and $K_L \to \pi^+\pi^-$ channels have suggested a tiny but nonzero direct weak CP-violating signal that is consistent with predictions of the KM model [5]. However, it is important to note that various ideas have been proposed in addition to the KM picture to account for the observed T-violation. Three representative examples are:

1. *Higgs boson models.* Weinberg and Lee have noted that if we introduce not just a single Higgs doublet but at least three such doublets, it is possible for CP to be violated in the exchange of charged Higgs scalars between quarks, giving rise to an effective "current"–"current" model of T-violation [6], where the "currents" have the form of scalar and pseudoscalar densities because the Higgs itself is a scalar:

$$\text{“current”} \sim \bar{q}'(1 \pm \gamma_5)qm_q. \tag{5.27}$$

In such a model, the scale of the "current" coupling is set by the mass of the quark or lepton to which the $H^\pm$ couples. Then,

$$\frac{H_{\text{Higgs}}}{H_{\text{strong}}} \sim \alpha_{em} \times \frac{m_q m_{q'}}{m_H{}^2}, \tag{5.28}$$

where m_H is the (unknown) Higgs mass. As mentioned earlier, no experimental evidence for the existence of either a charged or neutral Higgs has yet been reported. However, bounds *have* been placed on m_H at both the low and high ends, though a generous range, ~ 10 GeV–~ 300 GeV, is allowed [7].

2. *Left-right models.* Earlier we noted that a number of authors have proposed models in which the electroweak interaction is mediated by gauge bosons coupling both left- *and* right-handedly [8]. The fact that the right-handed boson is more massive accounts for the predominantly left-handed structure observed in low energy weak processes. Now, in general, the coupling of the right-handed gauge particle will occur with its own set of mixing parameters. Thus, although some models postulate the identity of the left- and right-handed KM matrices

$$U_L = U_R \tag{5.29}$$

in more general models, there is complete independence between U_L and U_R and, in fact, the additional freedom associated with the right-handed current even permits possible T-violating phases to appear in the light quark sector, in contrast to the KM model, wherein T-violation is associated only with the heavy quarks. [9].

3. *Superweak model.* This idea (not really a model) was introduced by Lincoln Wolfenstein shortly after the discovery of CP violation in 1964 [10]. He postulated the existence of a *very* weak (superweak) interaction that can change strangeness by two units

$$\langle \bar{K}^0 | H(\text{superweak}) | K^0 \rangle \neq 0. \tag{5.30}$$

Because of its weakness, the effects of such an interaction would only be visible because of the extraordinary sensitivity available in the $K^0,\bar{K}^0$ system. It follows that no effects would be observable elsewhere, which is a somewhat discouraging prospect to experimentalist and theorist alike.

In principle, we can decide between such models provided evidence can be found for CP violation in other systems. For example, if *nothing* is seen outside $K^0 - \bar{K}^0$, then we might suspect the correctness of the superweak model. If, on the other hand, an effect is observed in nuclear beta decay, we might suspect a left-right model. If T-violation is observed only in heavy quark couplings, the KM picture might be correct.

The key to determining the validity of any of these models, however, is to seek these T-violating effects in many different systems. How does one do this? Here are several possibilities that have been pursued:

1. *Detailed balance.* The idea is to perform a scattering reaction both forward and in reverse. Take, for example (cf. Figure 5.2),

$$p + {}^{27}\text{Al} \leftrightarrow \alpha + {}^{24}\text{Mg}.$$

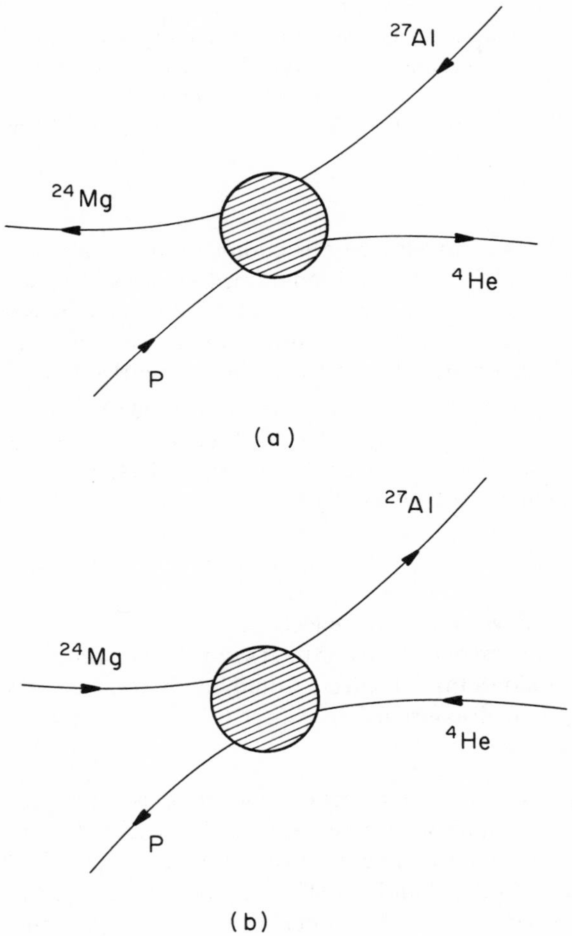

FIGURE 5.2. Time reversal invariance has been tested by looking for the equality of the differential cross sections (at corresponding scattering angles and center of mass energies) of the reactions $\alpha + {}^{24}\text{Mg} \leftrightarrow {}^{27}\text{Al} + p$.

Time reversal invariance requires that

$$\langle \beta_{p_2,s_2}|\mathrm{T}|\alpha_{p_1,s_1}\rangle = \langle \alpha_{-p_1,-s_1}|\mathrm{T}|\beta_{-p_2,-s_2}\rangle^*. \tag{5.31}$$

Since the differential scattering cross section is

$$\begin{aligned} &\text{Forward } \frac{d\sigma}{d\Omega} \sim \text{Kinematic terms} \times \sum_{\text{spins}} |\langle \beta|\mathrm{T}|\alpha\rangle|^2 \\ &\text{Reverse } \frac{d\sigma}{d\Omega} \sim \text{Kinematic terms} \times \sum_{\text{spins}} |\langle -\alpha|\mathrm{T}|-\beta\rangle|^2, \end{aligned} \tag{5.32}$$

we see that a precise relationship should exist between the two cross sections when measured at identical angles and energies in the center of mass. The best such measurements to date have been performed in the above system and have yielded [11]

$$\left|\frac{\text{Amp (T violating)}}{\text{Amp (T conserving)}}\right| < 5 \times 10^{-4} \tag{5.33}$$

which is an impressive confirmation of T-conservation in such reactions. However, *should* one have expected to see an effect? If time reversal arises at the level at which CP violation is observed, then since the scattering process involves the strong interaction, we might expect to observe a nonvanishing signal only at order

$$Gm_p{}^2 \times \varepsilon \sim 10^{-8}, \tag{5.34}$$

so that, to my mind, this is not a particularly sensitive test despite the obvious statistical precision.

2. *P-A comparison.* The goal here is to compare the polarization arising when, say, a proton is scattered at angle θ from an unpolarized target to the left-right asymmetry when polarized protons are scattered at the same angle from the same target. As shown diagrammatically in Figure 5.3, time invariance requires the equality

$$P = A. \tag{5.35}$$

Such an experiment has, for example, been performed at LAMPF. For np scattering at $p_L = 800$ MeV and $\theta = 133°$, we find [12]

$$P - A = 0.01 \pm 0.02. \tag{5.36}$$

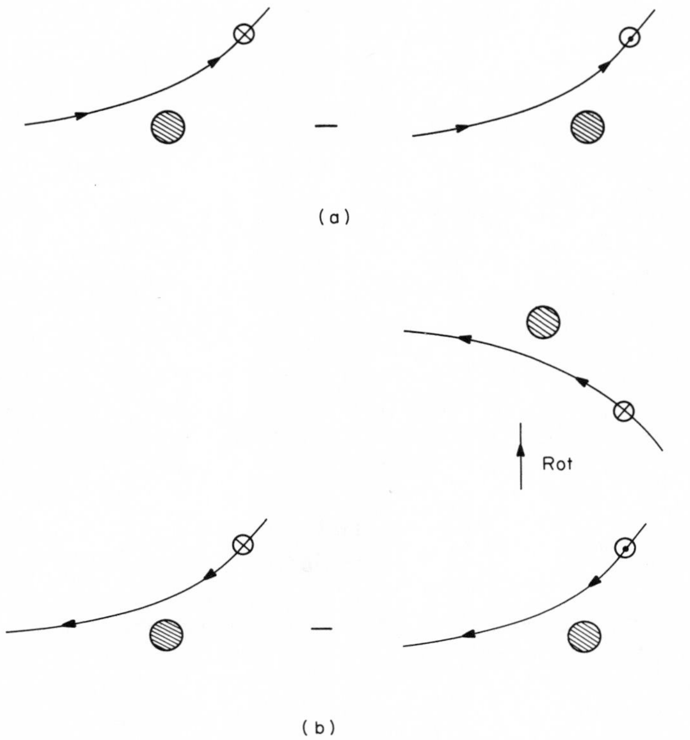

FIGURE 5.3. Time reversal invariance requires that the polarization produced when a particle is scattered from an unpolarized target (a) be equal to the left-right asymmetry in the scattering of a polarized beam from the same target (b).

However, again, since this scattering process involves the strong interaction, I again do not believe this to be a particularly sensitive laboratory for finding a possible T-violating effect. From my perspective, the most promising tests start with the weak or electromagnetic interaction, which involve the use of correlation experiments.

3. *Correlation experiments.* In order to see how such experiments can be used as probes for T-violation, consider the decay

$$\Lambda \rightarrow p\pi^-,$$

which occurs via the weak interaction, and thereby contains both a parity violating (*s*-wave) and parity conserving (*p*-wave) component.

The decay can be parameterized as

$$\langle p\pi^-|H_w|\Lambda\rangle = A_s\chi_p^\dagger\chi_\Lambda + A_p\chi_p^\dagger\boldsymbol{\sigma}\chi_\Lambda \cdot \hat{p}_{\pi^-}, \tag{5.37}$$

where χ are Pauli spinors for the respective baryons, and $\hat{p}_\pi$ is a unit vector in the direction of the outgoing pion. It is then straightforward to evaluate the decay spectrum

$$\begin{aligned} \frac{d\Gamma}{d\Omega_\pi} \sim 1 &+ A(\mathbf{J}_\Lambda + \mathbf{J}_p) \cdot \hat{p}_\pi + B\mathbf{J}_\Lambda \times \mathbf{J}_p \cdot \hat{p}_\pi \\ &+ C\mathbf{J}_\Lambda \cdot \mathbf{J}_p + (1 - C)\mathbf{J}_\Lambda \cdot \hat{p}_\pi \mathbf{J}_p \cdot \hat{p}_\pi, \end{aligned} \tag{5.38}$$

with $\mathbf{J}_\Lambda$,$\mathbf{J}_p$ being the lambda, proton polarization, respectively, and A,B,C being correlation functions that are given in terms of the weak decay amplitudes. For our purposes we focus on the so-called triple product correlation B. This spin-dependent term is apparently odd under time reversal, since

$$\mathbf{J}_\Lambda \times \mathbf{J}_p \cdot \hat{p}_\pi \rightarrow -\mathbf{J}_\Lambda \times -\mathbf{J}_p \cdot -\hat{p}_\pi. \tag{5.39}$$

Hence, in an experiment as shown in Figure 5.4, wherein a Λ polarized along the positive z direction emits a proton along the positive x-direction with polarization along the y-axis, T-invariance would seem to require $J_{Py} = 0$—i.e., the vanishing of the correlation coefficient B.

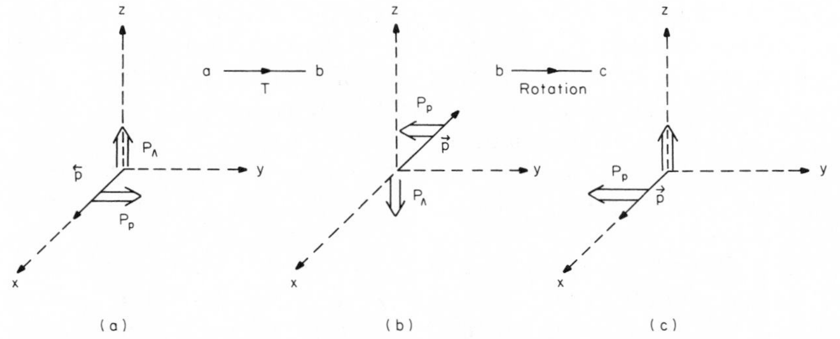

FIGURE 5.4. Shown in (a) is a Λ polarized along the z-axis decaying to a proton with polarization in the $+y$-direction. By time reversal and rotation invariance this appears to be equivalent to decay of the same Λ to a proton polarized along $-y$—(c). However, final state interactions must be taken into account, as discussed in the text.

Yet this is *not* the case! In fact, explicit calculation yields

$$B = \frac{2\mathrm{Im}A_s^*A_p}{|A_s|^2 + |A_p|^2}, \tag{5.40}$$

where, by the Fermi-Watson theorem [13]

$$\begin{aligned} A_s &= |A_s| \exp(i\delta_s) \\ A_p &= |A_p| \exp(i\delta_p), \end{aligned} \tag{5.41}$$

with δ_s, δ_p being the *s,p*-wave $\pi^- p$ phase shifts at a center of mass energy equal to the lambda mass. We find then $B \neq 0$. What did we do wrong? The problem has arisen because we have not *actually* run the reaction backwards, which would involve $p\pi^- \to \Lambda$ and cannot be done. That is, strictly speaking, T-invariance yields the result

$$\langle \psi_{\pi^- p}(\mathrm{out})|H|\Lambda\rangle = \langle \Lambda^T|H|\psi^T_{\pi^- p}(\mathrm{in})\rangle^*, \tag{5.42}$$

where the superscript T indicates that we are dealing with the time-inverted state, with all spins and momenta reversed. A *true* time reversal test then requires not only reversing all spins and momenta but also interchanging initial and final states. The correlation discussed above is equivalent to the reversal of spins and momenta but obviously does not include interchange of initial and final states. Only to the extent that the amplitude in eq. (5.42) is hermitian does a nonvanishing value for the correlation $\mathbf{J}_\Lambda \times \mathbf{J}_p \cdot \mathbf{p}_\pi$ imply violation of time reversal. One could say (intuitively but somewhat imprecisely) that the triple product correlation vanishes at the instant of transition but becomes nonzero as the π^- and proton interact on their way to the detectors.

Thus, by trying to seek T-violation using triple product correlations, we must pay the price of final state interaction effects simulating bona fide T-nonconservation. Nevertheless, this is not a fatal problem, provided such effects are measurable or calculable. In the present situation the $\pi^- p$ phase shifts are known experimentally and T-violation can be sought by looking for any deviation from the above prediction—eqs. (5.40) and (5.41). The experiment was performed a number of years ago by Overseth and Roth [14] and did find a triple product correlation,

$$B = -0.10 \pm 0.07. \tag{5.43}$$

Analysis revealed, however, that the measured phase between s- and p-wave amplitudes,

$$\Delta^{\exp} = \arg \Delta_s - \arg \Delta_p = 9.0° \pm 5.5°, \tag{5.44}$$

is quite consistent with the difference

$$\delta_s - \delta_p = 6.5° \pm 1.5° \tag{5.45}$$

determined in low-energy $\pi^- p$ scattering [15], so that there is no evidence for T-violation in this process, although the error bars leave considerable room.

Although not strictly within the purview of the topic of weak interactions, it should be noted, for completeness, that experimentally one has also looked for correlations of the type

$$\mathbf{J} \cdot \mathbf{k} \times \mathbf{E}\mathbf{J} \cdot \mathbf{k}\mathbf{J} \cdot \mathbf{E} \tag{5.46}$$

between the third-rank alignment of a parent nucleus and the cross product of the photon momentum **k** and electric field **E** in the electromagnetic decay of an excited nuclear state. (The electric field vector is determined by using Compton scattering as a probe for linear photon polarization.) Recent experiments have been performed by the Cal Tech group on three different nuclear systems, and the results are presented in Table 5.1. Of particular interest is the case of ^{191}Ir wherein a nonzero effect is found at the fifteen standard deviation level [16]! Nevertheless this does *not* signify the presence of T-nonconservation. Here, as in the case of $\Lambda \to p\pi^-$, the final state interaction can simulate the existence of T-violation. In the

TABLE 5.1
EXPERIMENTAL AND THEORETICAL VALUES FOR THE T-ODD CORRELATION $J \cdot EJ \cdot kJ \cdot k \times E$

Nucleus	Exptl. Phase ($\times 10^4$)	Theor. Phase ($\times 10^4$)
^{57}Xe	-3.1 ± 6.5	-6
^{191}Ir	-47 ± 3	-43 ± 4
^{181}Xe	-12 ± 11	-1

Given are the experimental values [16] together with corresponding theoretical calculations of final-state interaction phases [17] for the T-odd correlation $\mathbf{J} \cdot \mathbf{E}\mathbf{J} \cdot \mathbf{k}\mathbf{J} \cdot \mathbf{k} \times \mathbf{E}$.

present situation, these effects are small, being purely electromagnetic in origin and arising predominantly from the scattering of the outgoing gamma ray by an orbital electron. The calculated size of the final state effect [17] is shown in column two of Table 5.1. Obviously there is no evidence for a bona fide T-violating signal. However, since the basic transition element is electromagnetic, one might not expect to see anything until

$$\mathcal{O}\left(\frac{Gm_p{}^2\varepsilon}{\alpha}\right) \sim 10^{-6}.$$

The most sensitive triple product correlation tests involving the weak interaction have been performed using nuclear beta decay, for which one finds the decay spectrum

$$\frac{d\Gamma}{d\Omega_e d\Omega_\nu} \sim 1 + A\mathbf{J}\cdot\mathbf{p}_\beta/E_\beta + \ldots + D\mathbf{J}\cdot\mathbf{p}_\beta \times \hat{p}_\nu/E_\beta + \ldots + R\mathbf{J}\cdot\mathbf{p}_\beta \times \boldsymbol{\sigma}/E_\beta, \tag{5.47}$$

when $\mathbf{J},\boldsymbol{\sigma}$ signify the nuclear, beta polarizations, respectively. The triple product correlation D is sensitive primarily to a possible phase difference between Fermi and Gamow-Teller couplings,

$$D \sim \frac{2\,\mathrm{Im}\,a^*c}{|a|^2 + |c|^2} + \ldots, \tag{5.48}$$

and has the interesting property that the final state interaction effect vanishes to lowest order (for analog decay the leading final state term involves interference between the Gamow-Teller and weak magnetism terms [18] and is thus correspondingly tiny—$\sim 10^{-4}$). The experimental observable involves the correlation between the polarization of the parent nucleus and the cross product of the beta and neutrino momenta. Such experiments are exceedingly challenging because the neutrino is, strictly speaking, unobservable. Hence we must infer its momentum by detecting the recoil momentum imparted to the daughter nucleus by the decay. Nevertheless, very precise measurements have been carried out for both neutron beta decay and for the positron decay of ^{19}Ne, as shown in Table 5.2. Also indicated are the calculated electromagnetic final state effects. In the case of ^{19}Ne we are nearly at the level of sensitivity wherein we are limited by the final state interaction. In fact it would be of real interest to observe a signal at the calculated level, since this could then be used as a test of weak magnetism! Using my above estimates, we can see that

TABLE 5.2
EXPERIMENTAL AND THEORETICAL VALUES FOR THE T-ODD CORRELATION $J \cdot p_e \times p_\nu$

Decay	D^{expt}	$D^{\text{theor}}(\text{FSI})$
$^{19}\text{Ne} \to {}^{19}\text{F} + e^+ + \nu_e$	$(0.7 \pm 6) \times 10^{-4}$	2.5×10^{-4}
$n \to p + e^- + \bar{\nu}_e$	$(-1.1 \pm 1.7) \times 10^{-3}$	1.1×10^{-5}
	$(2.2 \pm 3.0) \times 10^{-3}$	

Given are the experimental values [19] together with corresponding theoretical calculations of final-state interaction effects [18] for the T-odd correlation $\mathbf{J} \cdot \mathbf{p}_e \times \mathbf{p}_\nu$.

these beta-decay triple product correlation experiments are *already* at a significant level of sensitivity, and the lack of any effect is trying to tell us something—but what?

The second triple product correlation coefficient R is also of interest since it can be shown to be primarily sensitive to the imaginary part of an interference between the leading V,A terms and any possible scalar, pseudoscalar, or tensor interaction pieces that might be present in nuclear beta decay. While this may sound exotic, the existence of a light-charged Higgs meson could lead to such terms. In the case of R, however, in contradistinction to D, the electromagnetic final state interaction is calculated to be quite large [20]:

$$R_{\varepsilon m} \approx \frac{\alpha Z m_e}{p_\beta} A, \tag{5.49}$$

where A is the beta asymmetry parameter in eq. (5.47). Detection of a bona fide T-violating effect requires a significant deviation from eq. (5.49). The first step in such a program was the measurement at Princeton of $R(^{19}\text{Ne})$. The nucleus ^{19}Ne was used because

$$A_{\text{exp}}(^{19}\text{Ne}) \approx -0.04, \tag{5.50}$$

which minimizes any possible final-state similation ($R_{\varepsilon m}^{\text{theo}} \sim 10^{-3}$). The experimental results [21]

$$R^{\text{exp}} = -0.079 \pm 0.053 \tag{5.51}$$

were not precise enough to make a significant test of T-violation. However, they indicate the potential promise of this technique.

The most sensitive tests of time reversal invariance yet performed involve the search for a neutron electric dipole moment. We note that under time reversal, one has for possible $E1,M1$ interactions:

$$\begin{aligned} M1 \qquad & U_M = -d_M\mathbf{S}\cdot\mathbf{B} \underset{T}{\longrightarrow} -d_M(-\mathbf{S})\cdot(-\mathbf{B}) = U_M \\ E1 \qquad & U_E = -d_E\mathbf{S}\cdot\mathbf{E} \underset{T}{\longrightarrow} -d_E(-\mathbf{S})\cdot\mathbf{E} = -U_E. \end{aligned} \tag{5.52}$$

Hence a nonvanishing electric dipole moment ($d_E \neq 0$) violates T-invariance. Such a term also violates parity. To see this, consider an electric field reflected in a mirror. If the field is produced by a set of capacitor plates, we have a situation as shown in Figure 5.5. Hence

$$U_E = -d_E\mathbf{S}\cdot\mathbf{E} \underset{P}{\longrightarrow} -d_E\mathbf{S}\cdot(-\mathbf{E}) = -U_E, \tag{5.53}$$

as claimed. Because of these symmetry restrictions, we expect any electric dipole moment to be small. Thus, the size of the magnetic dipole moment, which is *allowed* by parity and time reversal, is given by

$$d_M \sim \int d^3r\mathbf{r}\times\mathbf{j} \sim QR \tag{5.54}$$

or in terms of *ecm*, since $R \sim 1\text{ fm} = 10^{-13}$ cm

$$d_M \sim 10^{-13}\text{ ecm}. \tag{5.55}$$

An electric dipole moment should then be

$$\begin{aligned} d_E \sim d_M \times (Gm_p^2)\,&(\text{weak parity violation}) \\ &\times \varepsilon(\text{CP-violation}) \sim 10^{-21}\text{ ecm}, \end{aligned} \tag{5.56}$$

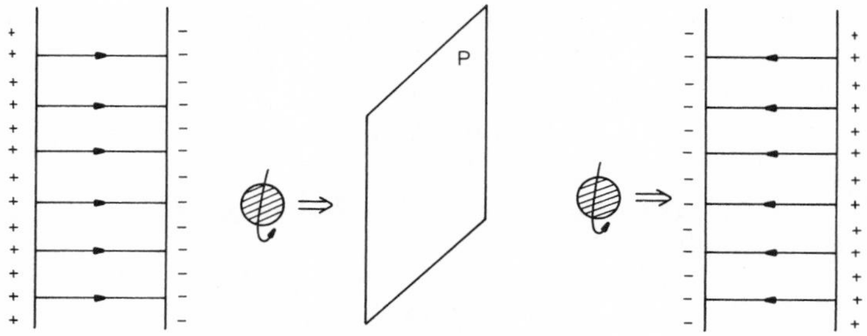

FIGURE 5.5. As reflected in a mirror, the spin parallel to an electric field on the left of the mirror becomes antiparallel on the right. Hence by P-invariance an interaction of the form $\mathbf{S}\cdot\mathbf{E}$ is forbidden.

already a *very* small number. However, experimentally the limit [22]

$$d_E < 2 \times 10^{-25} \text{ ecm} \tag{5.57}$$

has been placed. What does this say about the validity of various models of CP-violation?

Obviously, this result is consistent with the superweak model, which essentially predicts zero. It is also consistent with the KM model, which states that any T-violation must be associated with an interaction involving heavy quarks. In the case of the neutron, this means that d_E must involve the weak interaction at least in *second* order, as shown, for example, in Figure 5.6, giving an effect which has been estimated as [23]

$$d_E \sim 10^{-32} \text{ ecm}. \tag{5.58}$$

The KM model is thus in agreement with the experimental measurement, as is the left-right model, wherein a natural size of $\sim 10^{-26}$ ecm is expected [24].

The Higgs-exchange model is, however, challenged, as the natural scale of such a model can be shown to be

$$d_E \lesssim 10^{-25} \text{ ecm}. \tag{5.59}$$

For this reason, were the experimental upper bound to be lowered by an order of magnitude, the Higgs picture would probably be ruled out [25].

There is one more source of CP-violation that is important to note at this point. A careful mathematical treatment of QCD reveals that in addi-

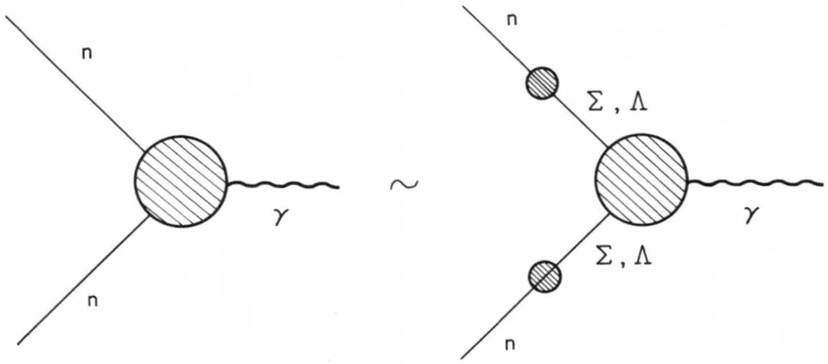

FIGURE 5.6. A typical contribution to the neutron electric dipole moment is this pole diagram wherein the photon interacts with virtual Λ,Σ° hyperons. Note that the weak interaction occurs twice here.

tion to the obvious terms, there exists an additional piece of the Lagrangian, called the anomaly.

Without going into detail, an "anomalous" component of the Lagrangian arises when a symmetry that is present in the classical theory is broken upon quantization [26]. In our case the symmetries are parity and time reversal, and the anomaly has the form

$$\mathscr{L}_{\mathrm{QCD}} = \ldots + \frac{\theta}{16\pi^2} F_a^{\mu\nu} \varepsilon_{\mu\nu\alpha\beta} F_a^{\alpha\beta}, \tag{5.60}$$

where θ is an arbitrary coefficient [27]. Since $\mathscr{L}_{\mathrm{QCD}}$ depends on the scalar product of the color electric and magnetic fields

$$F_a^{\mu\nu} \varepsilon_{\mu\nu\alpha\beta} F_a^{\alpha\beta} \sim \mathbf{E}^c \cdot \mathbf{B}^c, \tag{5.61}$$

we note that

$$\mathscr{L} \begin{array}{l} \xrightarrow{T} \mathbf{E} \cdot (-\mathbf{B}) = -\mathscr{L} \\ \xrightarrow{P} (-\mathbf{E}) \cdot \mathbf{B} = -\mathscr{L}, \end{array} \tag{5.62}$$

so that the anomalous piece of $\mathscr{L}_{\mathrm{QCD}}$ is, as claimed, odd under *both* parity and time reversal and, hence, can lead to $d_E \neq 0$. The electric dipole moment expected in such a model can be estimated by using the feature that this Lagrangian has a well-defined coupling of the pion to nucleons [28]

$$\mathscr{L}_{\mathrm{eff}} = \bar{g}_{\pi NN} \bar{N} \boldsymbol{\tau} N \cdot \boldsymbol{\phi}_\pi, \tag{5.63}$$

where $\bar{g}_{\pi NN}$ can be calculated using chiral symmetry. The neutron dipole moment then arises, for example, out of the diagram in Figure 5.7, yielding [29]

$$d_E(n) \sim e \frac{g_{\pi NN} \bar{g}_{\pi NN}}{4\pi^2 M_N} \ln \frac{M_N}{m_\pi} \sim 4 \times 10^{-16}\, \theta \text{ ecm}, \tag{5.64}$$

which then together with the experimental bound implies

$$\theta \lesssim 10^{-10}. \tag{5.65}$$

This result has at least two interesting implications. One is phenomenological. Henley and Haxton have noticed that the size of the time-reversal violating effect can be enhanced considerably within the nuclear medium [30]. Namely, if a nucleon emits a pion which is then absorbed by a second nucleon (cf. Figure 5.8) a large nuclear dipole moment may arise

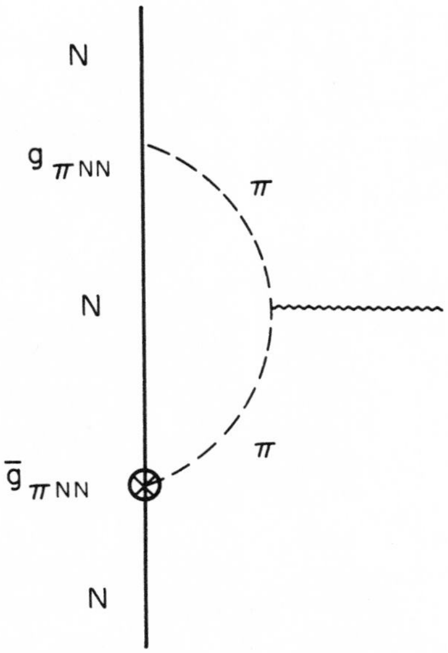

FIGURE 5.7. In the θ-model, a neutron electric dipole moment can arise from a simple pion exchange diagram, wherein one vertex is strong and CP, P conserving and the other, arising from the anomaly, is CP and P violating.

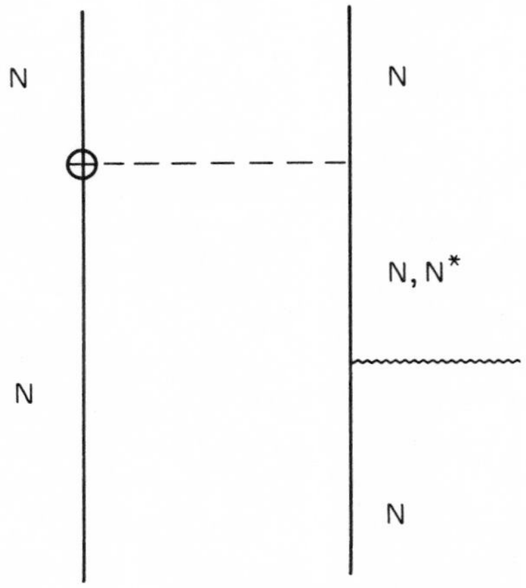

FIGURE 5.8. Haxton and Henley have shown that large enhancement affects can occur in the θ-model when the simple γ-nucleon interaction is accompanied by pion exchange between nucleons.

if nearby states of opposite parity are present. Specific examples considered by these authors include

$$\frac{d_E(\text{nucleus})}{d_E(\text{nucleon})} \sim \begin{cases} 10 & {}^{161}\text{Dy} \\ 100 & {}^{153}\text{Sm.} \\ 1000 & {}^{224}\text{Pa} \end{cases} \tag{5.66}$$

This nuclear "amplification" implies that in such a model a *nucleon* electric dipole moment at the 10^{-26} ecm level would imply a *nuclear* dipole moment in ^{153}Sm at the 10^{-24} ecm level. Of course, these calculations are within a specific (anomaly) model. However, the phenomenon of nuclear enhancement may be more general. Indeed, Sushkov et al. have asserted that the existence of kaon exchange between nucleons allows the nuclear electric dipole moment in the standard (KM) model to be nearly two orders of magnitude larger than its nucleon counterpart [31]. While this estimate is probably somewhat optimistic [32], it is clear that measurement of such *nuclear* electric dipole moments should be a component of a systematic attack on the source of T-violation.

At the experimental level, of course, one does not measure the nuclear moment directly. Rather what *is* measured is the electric dipole moment of a neutral atom. Such programs are presently underway at the University of Washington, and at Amherst and Yale. The best existing numbers at the present time are given for Xe and Hg by the Seattle group [33]

$$\begin{aligned} d_E(^{129}\text{Xe}) &= (0.2 \pm 1.1) \times 10^{-26}\ \text{ecm}, \\ d_E(^{199}\text{Hg}) &= (0.7 \pm 1.5) \times 10^{-26}\ \text{ecm} \end{aligned} \tag{5.67}$$

and improvement by at least two orders of magnitude is contemplated [34]. Unfortunately, interpretation of such *atomic* measurements in terms of an effective *nuclear* edm is not completely straightforward. For one thing, the electron itself may have a nonzero electric dipole moment. Secondly, even in models such as KM or "θ" wherein the electron edm vanishes, the full effect of the nuclear moment is screened due to Schiff's theorem, which states that in the limit of a point nucleus and a nonrelativistic model for the electron the effect of an externally applied electric field must vanish at the location of the nucleus [35]. Because the electron is not truly nonrelativistic and since realistic nuclei have a finite size, there does exist some sensitivity to the applied field. However, calculations by Henley and Haxton reveal that the effect is suppressed by a (model-dependent) factor of 10^3 or so [30]. Nevertheless, improvements in experimental methods may make such nuclear edm measurements competitive with the ongoing neutron experiments.

Returning to the CP violating anomaly, we note that this term has an additional, more fundamental significance. The coefficient θ is not restricted by any sort of symmetry or other constraints. Yet it is extraordinarily small—$\theta < 10^{-10}$. Why? Of course, it could just be accidental. However, this seems unlikely, especially since even if θ were given some tiny value in the fundamental Lagrangian, it would be renormalized by weak interaction effects. To be more specific, the weak interactions will in general generate a quark mass matrix,

$$\mathscr{L}_{\text{mass}}^{\text{wk}} = \bar{Q}_L M Q_R \quad \text{with } Q = \begin{pmatrix} u \\ d \\ s \end{pmatrix}, \tag{5.68}$$

which by a redefinition of the fields [i.e., a chiral SU(3) ⊗ SU(3) rotation] can be made diagonal and real *up to a phase* without affecting the rest of the QCD Lagrangian. The phase can only be removed, however, at the cost of changing the coefficient of the anomaly term

$$\theta \to \theta_0 + \delta\theta = \theta_{\text{eff}}. \tag{5.69}$$

Thus even if $\theta_0 = 0$, in general $\theta_{\text{eff}} \neq 0$ and it is even *more* peculiar why the effective θ is so tiny. A possible way out of this dilemma was posed by Peccei and Quinn [36], who suggested the existence of an additional symmetry in the QCD Lagrangian. The precise way in which this symmetry is imposed is rather technical, and we shall not discuss it here. However, let us note a phenomenological implication of this proposal, which is that there should exist a light pseudoscalar meson, the axion. (We can think of this particle arising as a Goldstone boson associated with the spontaneous breaking of the U(1) symmetry which multiplies all the left- or right-handed quarks by a common phase.) The properties of such a particle are completely predicted in this approach [37]:

$$\text{(i) Mass } m_a{}^2 = \frac{F_\pi{}^2}{F_s{}^2} m_\pi{}^2 \frac{m_u m_d}{(m_u + m_d)^2} \tag{5.70}$$

$$\text{(ii) Coupling to Electrons } \mathscr{L}_{\text{eff}} = -\frac{a}{F_s} m_e i\bar{e}\gamma_5 e \tag{5.71}$$

etc.,

where F_s is an unknown axion coupling constant. (We also note that the CP violating coupling of the π^0 is predicted in this approach as

$$\mathscr{L}_{\text{eff}} = -\frac{\theta_{\text{eff}}}{F_\pi}\frac{m_u m_d}{m_u + m_d}\pi^0(\bar{u}u - \bar{d}d), \tag{5.72}$$

which gives rise to the electric dipole moment estimates quoted above.)

Many searches have been made for such axions, but to date none have hit paydirt. Examples include:

1. Search for $K^+ \to \pi^+ a$ [38]. Unfortunately, this experiment is limited by serious backgrounds from $K^+ \to \pi^+\pi^0$ and $K^+ \to \mu^+\nu_\mu$. Nevertheless, the absence of an axion-associated peak indicates $F_s > 10^5$ GeV.
2. Nuclear experiments have examined the decays of the 15.11 MeV, 1^+ and 12.7 MeV 1^+ of ^{12}C for evidence of the mode

$$^{12}\text{C}^*\,(15.11\ \text{MeV}, 1^+) \to {}^{12}\text{C} + a \searrow e^+e^-,$$

which is predicted to occur with a branching ratio of 10^{-5} or so [39]. Experiments have thusfar been negative, ruling out standard axions in the mass range of 2 – 12 MeV, etc. Nevertheless, modifications of the simple Peccei-Quinn scheme which *are* consistent with present experimental limits have been proposed, and the search for axions continues [40]. Further discussion, however, would take us too far afield.

Summarizing then, nuclei have played and doubtless will continue to play an important role in the experimental search for the origin of CP/T violation. Triple product correlation measurements will continue to be improved. The most promising avenue appears to be associated with the search for the existence of a nuclear electric dipole moment. A second program, which looks almost equally sensitive, is to seek the existence of a T- and P-violating term

$$\mathbf{s}_n \cdot \mathbf{J} \times \mathbf{k}_n \tag{5.73}$$

in the index of refraction of a polarized neutron with spin $\mathbf{s}_n$, momentum $\mathbf{p}_n$ propagating through matter having nuclear polarization $\mathbf{J}$ [41]. By choosing materials such as ^{139}La wherein a p-wave resonance occurs in the compound nucleus very close to threshold, we can maximize the P-odd, T-even effect [42] and thereby gain the most sensitivity to any possible T-odd term that might be present. Such an experiment is being conducted at present by a Washington/Los Alamos/Princeton collaboration [43]. However, this is a challenging task and results will not soon be forthcoming.

Having given an overview of the present situation in the nuclear T-violation arena, we move on to the subject of parity noninvariance.

5.2 Nuclear Parity Violation

Above we mentioned the important feature that the weak interaction, unlike its electromagnetic and strong counterparts, has a definite handedness and therefore violates parity. That nature should prefer a definite handedness came as quite a shock to physicists in 1957 in spite of the fact that our culture, and indeed our biology, *does* make a definite choice. Thus most individuals in our society are right-handed (although among theoretical physicists personal observation would put the fraction at only slightly above one-half) and one might even say that our culture is "dexist" in this regard. Indeed, when we compare the words for left and right in various languages and their corresponding English derivatives,

	left	*right*
Latin	sinistra	dextra
English	sinister	dextrous
French	gauche	droit
English	gauche	adroit
	etc.,	

we can see a definite prejudice against "southpaws." However, this clearly has nothing to do with the laws of physics.

Another curious feature in this regard has to do with the structure of amino acids. Consider a generic amino acid whose structure is shown in Figure 5.9 [44]. Nearly all such acids occur in two different forms, each

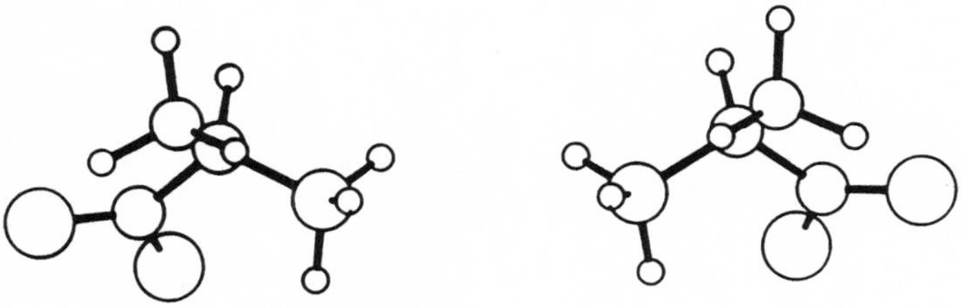

FIGURE 5.9. Shown are models of *L*- and *D*-type amino acids which rotate the plane of polarization of a light beam to the left and right, respectively.

having a definite handedness. In fact, this handedness can be observed at the macroscopic as well as the microscopic level. A solution of right-handed molecules will rotate the electric field vector of plane-polarized light to the right, while a solution of left-handed molecules will rotate the plane of polarization to the left (cf. Figure 5.10). It is remarkable, then, that *all* such molecules found in living proteins are of the left-handed variety. (As Feynman has emphasized [45], a frog consisting purely of right-handed acids could in principle be constructed. However, if placed in a lily pond, the creature would soon expire, as its system would be unable to digest the left-handed molecules found in all plants/insects present in this ecosystem.) On the other hand, there is nothing particularly special about the left-handed species. In fact, chemically synthesized amino acids consist of a statistical mixture—50% left-handed and 50% right-handed.

Why then are all biologically occurring acids of the left-handed variety? One possibility is that when life formed on the earth, it may have evolved outward from some sort of accidental occurrence at a *single* location. Of course, that accidental happening had a 50% chance of involving left-handed species and a 50% chance of involving right-handed species, so there was nothing predestined in the left-handed choice. Rather, once this "decision" was made, all life evolved from this point and was genetically destined to continue with the same handedness. Had life evolved more or less simultaneously from many points upon the earth's surface, we might expect that present biology should consist of some statistical mixture of left- and right-handed varieties. However, these are only speculations.

The point I have tried to make in the previous arguments is that the strictly left-handed choice made by terrestrial biology does *not* involve any fundamental law of physics. Perhaps the most vivid picture of the *physics* issues associated with parity nonconservation is the one mentioned earlier by Feynman [46], who asks us to imagine that communication

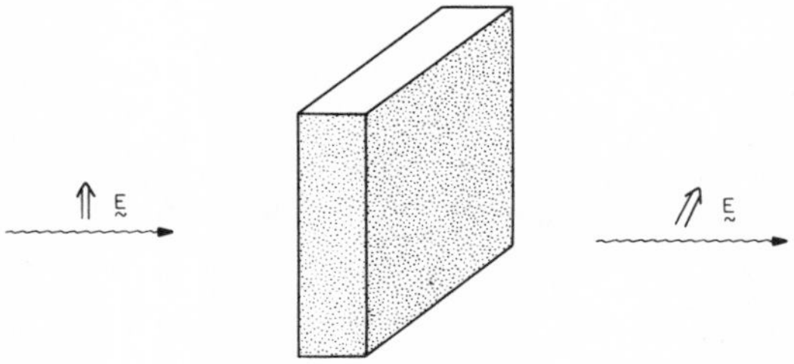

FIGURE 5.10. After passage through a solution containing molecules of predominantly a single handedness, the plane of polarized light is rotated, as shown.

has finally been established with some sort of extraterrestrial civilization. This communication is via radio signals and, of course, we can assume that many eons will pass until we have learned one another's languages well enough to communicate easily. Nevertheless, at some point—maybe when we are attempting to explain on which side of the body our heart is placed—it becomes necessary to define the concept of "left" and "right." Here, then, is our problem. Can the extraterrestrial perform some experiment that will favor one particular handedness? (It is *not* fair game here to transmit a circularly polarized signal; an experiment that can be performed solely at one end is required here.) If parity were a conserved quantity, such an experiment would not be possible and this is what nearly all physicists believed (primarily on esthetic grounds) until developments during the mid-1950s proved otherwise. The discovery of parity nonconservation associated with the weak interaction revealed that such a measurement does exist—e.g., the experiment of Wu et al. [47] which demonstrated that electrons from the beta decay of polarized ^{60}Co tended to be emitted antiparallel to the direction of nuclear spin. Since parity invariance would require mirror image configurations to be equally likely, we see from Figure 5.11 that preferred emission antiparallel to the spin orientation clearly indicates a violation of parity. Thus we *can* communicate to an extraterrestrial friend a measurement that will be able to distinguish left from right. (Of course, if CP invariance were valid and the extraterrestrial were actually an anti-extraterrestrial doing experiments with anti-^{60}Co, then left and right would still be confused, but that is another matter.)

But enough of this digression and back to the problem at hand: we have discussed the way in which parity noninvariance is observed in the regime

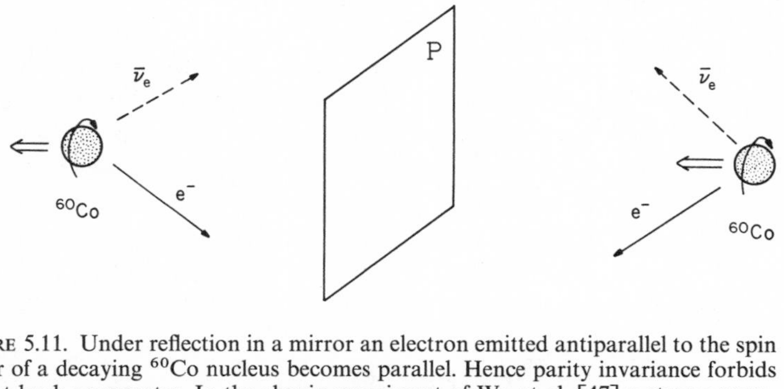

FIGURE 5.11. Under reflection in a mirror an electron emitted antiparallel to the spin vector of a decaying ^{60}Co nucleus becomes parallel. Hence parity invariance forbids a front-back asymmetry. In the classic experiment of Wu et al. [47] a strong asymmetry was discovered, indicating violation of parity by the weak interaction.

of nuclear beta decay. However, parity violation is also manifested in the so-called nonleptonic sector—that is, in interactions without leptons or neutrinos—where it appears as a small contaminent to the basic nucleon-nucleon force [48].

To see how this comes about, we recall the form of the Cabibbo (charged) current

$$\frac{1}{2}J_\mu{}^W = \frac{1}{2}J_\mu^{\text{lepton}} + \cos\theta_c \bar{u}_L\gamma_\mu d_L + \sin\theta\bar{u}_L\gamma_\mu s_L + \cdots \tag{5.74}$$

Thus, using only the simple Fermi interaction,

$$H_w = \frac{G}{\sqrt{2}} J_\mu{}^W J_\mu^{W\dagger}, \tag{5.75}$$

we find a piece of the weak Hamiltonian that involves only quarks, carries zero strangeness, and contains a parity-violating component,

$$H_w^{NL}(\Delta S = 0) = \frac{4G}{\sqrt{2}}(\cos^2\theta_c \bar{d}_L\gamma_\mu u_L \bar{u}_L\gamma^\mu d_L + \sin^2\theta_c \bar{s}_L\gamma_\mu u_L \bar{u}_L\gamma^\mu s_L). \tag{5.76}$$

This weak force must be present in addition to the basic nucleon-nucleon interaction. The problem in observing such a component is that we are dealing with a *weak* interaction, which is therefore dwarfed by the much larger strong N–N force. The weak NN component is detectable only by the presence of parity-violation, which it alone possesses.

Now, the ordinary nucleon-nucleon potential can be represented, with reasonable quantitative precision, by a sum of single meson exchange diagrams [49], as shown in Figure 5.12a, where the sum is over π, η, ρ, ω, σ, etc., mesons that couple strongly to the $\bar{N}N$ vertex. In the same way we can attempt to represent the weak nucleon-nucleon force, except that in this case, one $\bar{N}NM$ vertex is strong, the other weak (cf. Figure 5.12b). A second difference between the weak and strong interactions allows somewhat of a simplification—according to CP invariance, exchange of a neutral spinless boson is forbidden. Hence, π^0, η^0, σ^0, ... exchange may be omitted. This result, derived under the conditions of hermiticity and CP invariance, usually goes under the name of "Barton's Theorem" [50].

One can estimate the size of this parity-violating interaction via

$$\frac{V_{NN}^{(-)}}{V_{NN}^{(+)}} \sim G m_N{}^2 \frac{p_F}{m_N} \sim 10^{-6}, \tag{5.77}$$

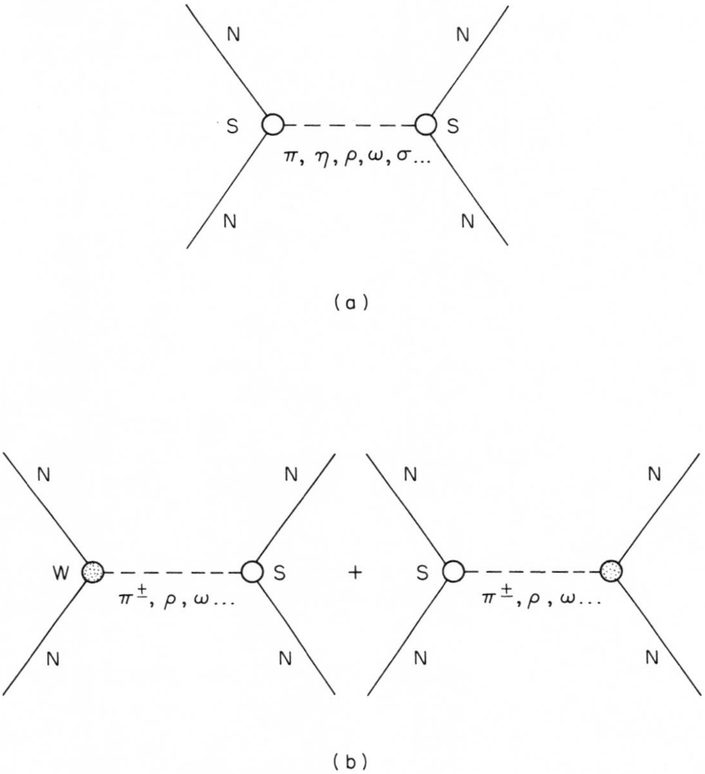

FIGURE 5.12. The usual nucleon-nucleon interaction $V_{NN}^{(+)}$ is reasonably well represented in terms of a single meson exchange picture, with light mesons π, η, ω, ρ, σ, etc., being involved (a). Likewise, the parity-violating interaction is viewed similarly except that π^0, η, σ exchanges are forbidden by Barton's theorem and one πNN vertex is weak and parity violating (b).

where $p_F \sim 250$ MeV is the Fermi momentum and is introduced by the requirement of having a parity *violating* observable.

As expected, this is a small effect and is consequently *very* challenging to detect. What sort of experiments can be utilized? I shall list three representative types, but there are many others.

1. *Longitudinal asymmetry measurements.* We argued in Chapter 2 that a parity transformation was equivalent to reflection in a mirror. Consider then the scattering of a longitudinally polarized proton from an unpolarized target as shown in Figure 5.13. Assuming parity invariance, the

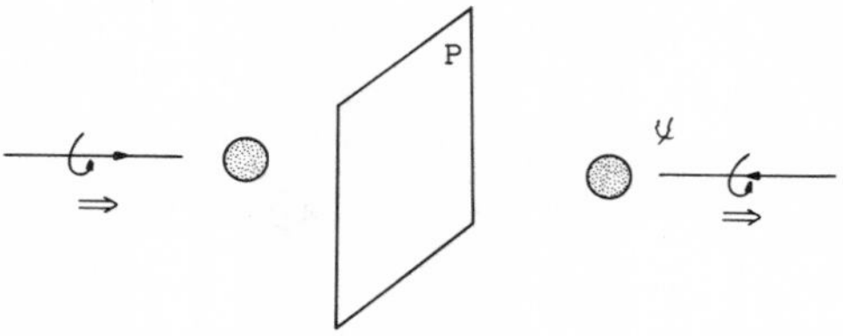

FIGURE 5.13. Under spatial inversion, a positive helicity state turns into its negative helicity analog. Hence parity invariance requires no helicity dependence in the scattering cross section.

processes on opposite sides of the mirror must occur with equal probability. Hence, the cross section for the scattering of left-handed protons must be equal to the cross section for right-handed protons. Typically, one defines the so-called asymmetry parameter A_L as the difference divided by the sum of the longitudinally polarized cross sections

$$A_L = \frac{\sigma_R - \sigma_L}{\sigma_R + \sigma_L}, \tag{5.78}$$

which must vanish in the absence of parity violation. A number of experiments of this type have been performed, typically involving a polarized proton beam on various targets

$$\begin{array}{rll} A_L(\vec{p} - p;\, 15\text{ MeV}) = & (-1.7 \pm 0.8) \times 10^{-7} & \text{LANL [51]} \\ A_L(\vec{p} - p;\, 45\text{ MeV}) = & (-1.50 \pm 0.22) \times 10^{-7} & \text{SIN [52]} \\ & (-1.3 \pm 2.3) \times 10^{-7} & \text{LBL [53]} \\ A_L(\vec{p} - d;\, 15\text{ MeV}) = & (-0.3 \pm 0.8) \times 10^{-7} & \text{LANL [54]} \\ A_L(\vec{p} - \alpha;\, 45\text{ MeV}) = & (-3.3 \pm 0.9) \times 10^{-7} & \text{SIN [55]} \\ A_L(\vec{p} - p);\, 800\text{ MeV}) = & (2.4 \pm 1.1) \times 10^{-7} & \text{LAMPF [56]} \\ A_L(\vec{p}, -H_20,\, 6\text{ GeV}) = & (2.6 \pm 0.6) \times 10^{-6} & \text{ANL [57]} \end{array} \tag{5.79}$$

2. A second category of experiment involves a nucleus in an excited state decaying via photon emission to its ground state. As shown in Figure 5.14, since reflection in a mirror reverses the circular polarization of normally directed photons, parity invariance requires emission of equal numbers of right- and left-handed circularly polarized photons. Hence, the existence of a nonzero circular polarization represents a probe for

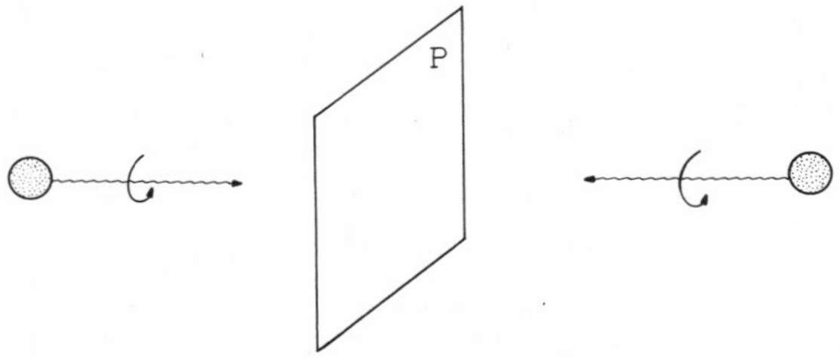

FIGURE 5.14. Under spatial inversion a right-hand circularly polarized photon becomes left-handed. Hence parity invariance forbids a net circular polarization.

parity violation. We can see this also in a more formal way: the operators for $E1$, $M1$ emission of radiation are, respectively,

$$\begin{aligned} &E1\colon \hat{\varepsilon}_\gamma \cdot \mathbf{p} \\ &M1\colon i\hat{\varepsilon}_\gamma \times \mathbf{q} \cdot \mathbf{r} \times \mathbf{p} = i\hat{\varepsilon}_\gamma \times \mathbf{q} \cdot \mathbf{L}\,. \end{aligned} \tag{5.78}$$

Circular polarization involves a linear combination of polarization states orthogonal to the photon momentum and 90° out of phase, i.e.,

$$\hat{k}_\gamma = \hat{z} \qquad \begin{aligned} \hat{\varepsilon}_R &= \frac{1}{\sqrt{2}}(\hat{x} + i\hat{y}) \\ \hat{\varepsilon}_L &= \frac{1}{\sqrt{2}}(\hat{x} - i\hat{y}). \end{aligned} \tag{5.81}$$

Since both **p** and **L** are tensors of rank one, the Wigner-Eckart theorem requires that $E1$ and $M1$ amplitudes are proportional to one another:

$$\langle f|\mathcal{O}_{E1}|i\rangle \propto \langle f|\mathcal{O}_{M1}|i\rangle. \tag{5.82}$$

Finally, since $\hat{\varepsilon}_\gamma \perp \hat{\varepsilon}_\gamma \times \hat{q} \perp \hat{q}$, we see that the simultaneous presence of both electric and magnetic dipole transitions *must* lead to circular polarization. However, the momentum operator **p** is a polar vector, $P\,\mathbf{p}\,P^{-1} = -\mathbf{p}$, while the angular momentum **L** is an axial vector, $P\,\mathbf{L}\,P^{-1} = \mathbf{L}$, leading to the selection rules

$$\begin{aligned} &E1\colon P_iP_f = -1 \\ &M1\colon P_iP_f = +1, \end{aligned} \tag{5.83}$$

so that clearly a violation of parity invariance is involved.

Several such experiments have been performed:

$np \to d\gamma$	$P_\gamma = (1.8 \pm 1.8) \times 10^{-7}$	Gatchina [58]
^{18}F(1.081 MeV)	$P_\gamma = (-7 \pm 20) \times 10^{-4}$	Cal Tech/Seattle [59]
	$(3 \pm 6) \times 10^{-4}$	Florence [60]
	$(-10 \pm 18) \times 10^{-4}$	Mainz [61]
	$(2 \pm 6) \times 10^{-4}$	Queens [62]
^{21}Ne(2.789 MeV)	$P_\gamma = (24 \pm 24) \times 10^{-4}$	Seattle/Chalk River [63]
	$(3 \pm 16) \times 10^{-4}$	Chalk River/Seattle [64]

(5.84)

3. A third technique examines the decay of a nuclear level that has been polarized, as shown in Figure 5.15. Equal probability of left- and right-side processes (i.e., parity conservation) then requires that equal numbers of decay products be emitted parallel and antiparallel to the direction of nuclear spin:

$$\text{Decay Prob}(\hat{k}_\gamma \cdot \mathbf{J} > 0) = \text{Decay Prob}(\hat{k}_\gamma \cdot \mathbf{J} < 0). \tag{5.85}$$

Analysis of the experiment is usually done in terms of the asymmetry parameter,

$$A_\gamma = \frac{N(\mathbf{J} \cdot \hat{k} > 0) - N(\mathbf{J} \cdot \hat{k} < 0)}{N(\mathbf{J} \cdot \hat{k} > 0) + N(\mathbf{J} \cdot \hat{k} < 0)}. \tag{5.86}$$

Such a measurement has been performed in ^{19}F with the result

^{19}F(110 KeV)	$A_\gamma = (-8.5 \pm 2.6) \times 10^{-5}$	Seattle [65]	(5.87)
	$(-6.8 \pm 1.8) \times 10^{-5}$	Mainz [66]	

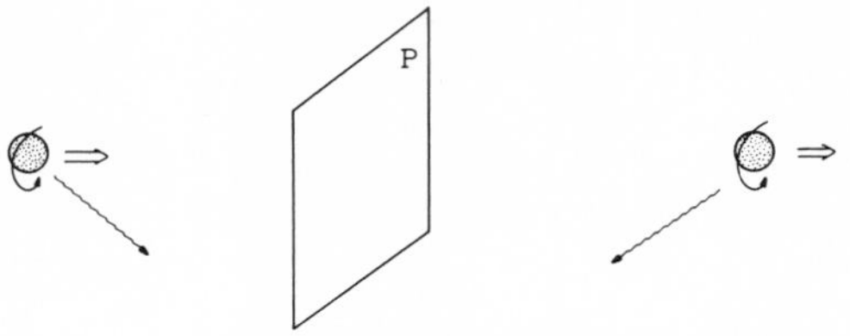

FIGURE 5.15. Under spatial inversion a photon emitted parallel to nuclear spin becomes antiparallel. Hence parity invariance forbids the existence of a front-back asymmetry.

and in ^{180}Hf yielding

$$A_\gamma = -(1.66 \pm 0.18) \times 10^{-2} \text{ [67].}$$

Finally, a similar experiment, performed at Grenoble, measures the asymmetry of the outgoing photon with respect to the direction of polarization of an incident thermal neutron beam in the reaction, $\vec{n} + p \to d + \gamma$, yielding an upper bound at the 10^{-7} level,

$$A_\gamma = -(4.7 \pm 4.7) \times 10^{-8} \quad \text{Grenoble [68].} \tag{5.88}$$

A quick look at these results indicates something peculiar, in that our earlier estimate indicated weak effects arising at the level 10^{-6}, and yet here we have some parity-violating observables detected at much higher levels, e.g., 10^{-5} in the case of ^{19}F and even 10^{-2} for ^{180}Hf. How can this be? The origin of this discrepancy is that we can use nuclei to amplify parity-violating signals. To see how this can be accomplished, consider a nucleus with an excited state of spin parity J^- which has a nearby level of the same spin but opposite parity J^+, both of which may decay down to a J'^+ ground state. Now, although I have labeled such states by their spin and parity, in reality none of these states is a true parity eigenstate because of the presence of the weak interaction. (Spin, of course, *is* a good quantum number because of angular momentum conservation.) In first-order perturbation theory, we can calculate the mixing of the close-by J^+,J^- levels, viz.,

J^+

J^-

J'^+

$$\begin{aligned} |\psi_{N^+}\rangle &\cong |\phi_{J^+}\rangle + \frac{|\phi_{J^-}\rangle\langle\phi_{J^-}|H_w|\phi_{J^+}\rangle}{E_+ - E_-} \\ &= |\phi_{J^+}\rangle + \varepsilon|\phi_{J^-}\rangle \\ |\psi_{J^-}\rangle &\cong |\phi_{J^-}\rangle + \frac{|\phi_{J^+}\rangle\langle\phi_{J^+}|H_w|\phi_{J^-}\rangle}{E_- - E_+} \\ &= |\phi_{J^-}\rangle - \varepsilon|\phi_{J^+}\rangle, \end{aligned} \tag{5.89}$$

in an obvious notation. Note that we have truncated the sum over *all* intermediate states down to a single state, by assuming these two levels to be very close in energy. (In principle, the ground state itself will contain a negative parity component also. However, this will be small assuming the absence of any nearby J'^- levels.) We can estimate the magnitude of the mixing by scaling to a typical nuclear energy splitting, of the order of an MeV or so. Then, associating this splitting with the strong interaction, we estimate

$$\langle \phi_{J^-}|H_w|\phi_{J^+}\rangle \sim \frac{H_w}{H_{st}} \times 1 \text{ MeV} \sim 1 \text{ eV}. \tag{5.90}$$

For two levels with a typical—MeV—spacing, we then have

$$\varepsilon = \frac{\langle \phi_{J^-}|H_w|\phi_{J^+}\rangle}{E_+ - E_-} \sim \frac{1 \text{ eV}}{1 \text{ MeV}} \sim 10^{-6}, \tag{5.91}$$

as expected. However, it is clearly advantageous to select nuclei wherein $|E_+ - E_-|$ is *very* small. An example is ^{21}Ne, where a 5.7 KeV splitting exists between $1/2^+$ and $1/2^-$ levels at 2.8 MeV (cf. Figure 5.16). Thus, we might expect a considerably enhanced mixing parameter:

$$\varepsilon \sim \frac{1 \text{ eV}}{6 \text{ KeV}} \sim 10^{-4}. \tag{5.92}$$

However, the situation is even better than this. The circular polarization is given by

$$\begin{aligned} &\frac{1}{2}^+ \text{ decay} \qquad P_\gamma \sim 2\varepsilon \frac{\langle E1\rangle}{\langle M1\rangle} \\ &\frac{1}{2}^- \text{ decay} \qquad P_\gamma \sim 2\varepsilon \frac{\langle M1\rangle}{\langle E1\rangle}, \end{aligned} \tag{5.93}$$

and in this transition, the $E1$ matrix element is enormously suppressed. A measure of this is the so-called $B(E1)$ value, which compares the measured rate of the transition to the value expected for an unsuppressed single particle transition. The measured value in ^{21}Ne for the $\frac{1}{2}^-$-ground state transition [69]

$$^{21}\text{Ne: } B(E1) \sim 1 \times 10^{-6} \text{ Weisskopf Units} \tag{5.94}$$

is much less than that found for typical $E1$ decays:

$$\langle B(E1)\rangle_{AV} \sim 10^{-1} \text{ Weisskopf Units.} \tag{5.95}$$

Thus we expect an additional enhancement of the order of a factor of about $\sqrt{10^{-5}} \sim 300$, and we choose the $\frac{1}{2}^-$ level in order to examine the

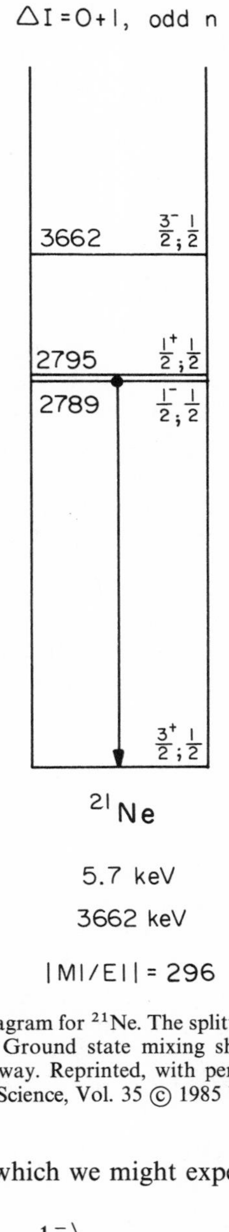

FIGURE 5.16. Shown is a level diagram for ^{21}Ne. The splitting between the nearby $\frac{1}{2}^+$ and $\frac{1}{2}^-$ levels is only 5.7 KeV. Ground state mixing should be negligible as the nearest $\frac{3}{2}^-$ level is 3.662 MeV away. Reprinted, with permission, from the Annual Review of Nuclear and Particle Science, Vol. 35 © 1985 by Annual Reviews, Inc.

circular polarization, for which we might expect

$$P_\gamma \sim \frac{2\left\langle \frac{1}{2}^+ \left| H_w \right| \frac{1}{2}^- \right\rangle}{5.7\ \text{KeV}} \times\cdot \sqrt{10^5} \sim \frac{\left\langle \frac{1}{2}^+ \left| H_w \right| \frac{1}{2}^- \right\rangle}{9\ \text{eV}}. \tag{5.96}$$

Detailed calculation yields the result [70]

$$\left|\frac{M1}{E1}\right| = 296 \tag{5.97}$$

and

$$P_\gamma\left(\frac{1}{2}^-\right) = \frac{\left\langle\frac{1}{2}^+ |H_w| \frac{1}{2}^-\right\rangle}{5.7\ \text{KeV}} \cdot 2\left|\frac{M1}{E1}\right| = \frac{\left\langle\frac{1}{2}^+ |H_w| \frac{1}{2}^-\right\rangle}{9.5\ \text{eV}}, \tag{5.98}$$

in agreement with our rough estimate in eq. (5.96).

Analysis of the other transitions proceeds in a similar fashion. In the case of ^{19}F, the mixing is between the $\frac{1}{2}^+$ ground state and the $\frac{1}{2}^-$ first excited state at 110 KeV (cf. Figure 5.17, left side) yielding [70]

$$\left|\frac{M1}{E1}\right| = 11 \quad \text{and} \quad A_\gamma = \left\langle\frac{1}{2}^- |H_w| \frac{1}{2}^+\right\rangle \Big/ 5200\ \text{eV} \tag{5.99}$$

For ^{18}F the splitting between the 0^+ excited state at 1.081 MeV (cf. Figure 5.17, right side) and the nearby 0^- level is only 39 KeV and we are faced with a strong $M1$ transition and relatively weak $E1$, yielding [70]

$$\left|\frac{M1}{E1}\right| = 112 \quad \text{and} \quad P_\gamma(1.08\ \text{MeV}) = \frac{\langle 0^+|H_w|0^-\rangle}{177\ \text{eV}}. \tag{5.100}$$

Below we shall utilize these results in order to deduce properties of the $\Delta S = 0$ nonleptonic Hamiltonian. However, for completeness it is important to note that there exists a rather substantial body of data involving parity-violating observables in heavy nuclei and on forbidden α-decays of light nuclei, which we shall not subject to theoretical scrutiny now. In the latter case, the width of the transition

$$^{16}\text{O}(2^-, 8.87\ \text{MeV}) \rightarrow {}^{12}\text{C} + \alpha,$$

for example, has been measured to be [71]

$$\Gamma_\alpha = (1.03 \pm 0.28) \times 10^{-10}\ \text{eV}. \tag{5.101}$$

However, while this result is roughly consistent with the value we might expect [72], a precise theoretical analysis is impossible, inasmuch as there

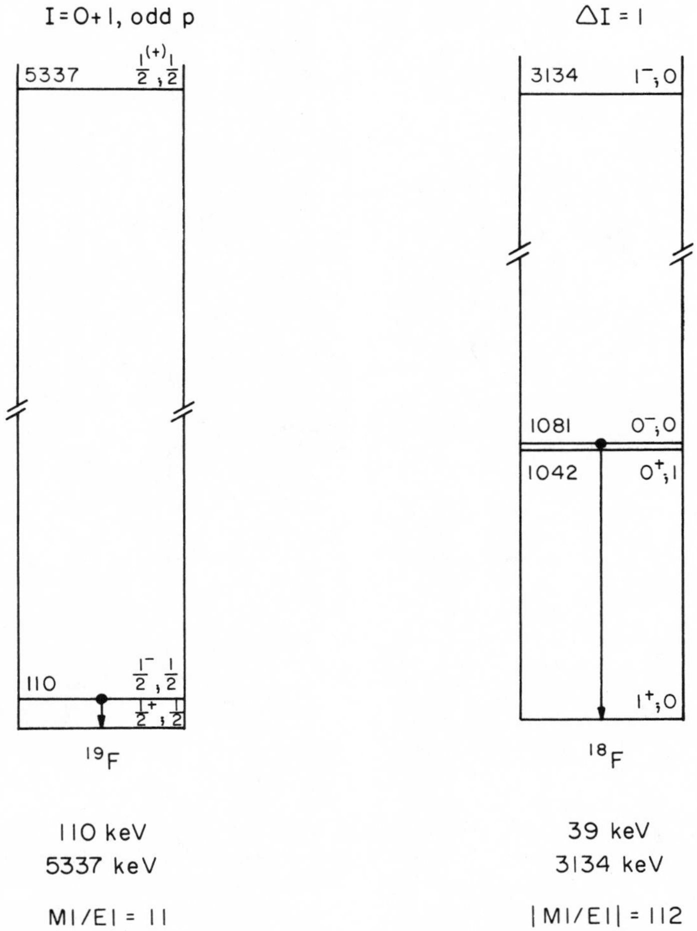

FIGURE 5.17. Shown are level diagrams for ^{19}F (a) and ^{18}F (b). The two-level approximation for ^{19}F should be quite adequate as the separation between the $\frac{1}{2}^+$ ground state and $\frac{1}{2}^-$ first excited state is only 110 KeV, while the nearest $\frac{1}{2}^+$ level is at 5.337 MeV. In the case of ^{18}F, the $0^+ - 0^-$ splitting is only 39 KeV, while the nearest 1^- state that could mix with the 1^+ ground state is at 3.124 MeV. Reprinted, with permission, from the Annual Review of Nuclear and Particle Science, Vol. 35 © 1985 by Annual Reviews, Inc.

is no single nearby 2^+ level in ^{16}O which is involved in the mixing. Rather, several 2^+ states exist in the vicinity which can be involved [cf. Figure 5.18]. Similar nuclear structure inadequacies plague the analysis of other light nuclear forbidden alpha decays [73].

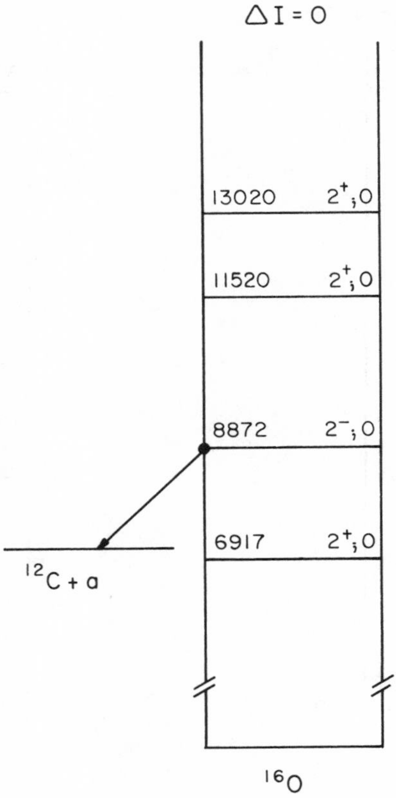

FIGURE 5.18. The simple two-level approximation is not adequate for the case of the 2^-, 8.87 MeV level of ^{16}O, as there are a number of nearby 2^+ levels. Reproduced, with permission, from the Annual Review of Nuclear and Particle Science, Vol. 35 © 1985 by Annual Reviews, Inc.

Likewise, even though good experimental results are available in systems such as $^{175}Lu(9/2^- \rightarrow 7/2^+)$, $^{181}Ta(5/2^+ \rightarrow 7/2^+)$, etc. [74], the inadequacy of the two-state approximation and the lack of reliable nuclear wave functions in this region indicate that, while such experiments are known to be roughly consistent with theoretical expectations [75], a more precise analysis is probably not warranted. In other heavy nuclear systems such as ^{180}Hf, where a 2% asymmetry has been observed in the 501-KeV photon arising from decay of the 8^- isomer [67] or in low energy neutron scattering from ^{139}La, where the longitudinal asymmetry of the total neutron cross section has been determined as [76]

$$A_L = (7.3 \pm 0.5) \times 10^{-2}, \tag{5.102}$$

even rough estimates of the expected size of the effect are suspect. In the case of ^{180}Hf, we are dealing with an enormously retarded transition having $\Delta K = 8$ in the Nilssen rotational model. In the case of ^{139}La, we have a narrow p-wave resonance sitting near a host of nearby s-wave resonances. In neither situation can we calculate reliably, and thus, although the size of such effects is spectacularly large, clearly revealing the presence of the parity-violating nuclear force, the use of these numbers to probe the structure of this interaction must be postponed.

As outlined above, many experiments have been performed seeking evidence for the existence of nuclear parity violation. What do they reveal about the structure of the nonleptonic weak Hamiltonian? To answer this question, we must calculate the weak mixing matrix elements theoretically. Many authors have attempted such evaluations [77] (with varying degrees of success). I choose to outline our group's work on this subject, since I am most familiar with it and it is representative of other work in the field [78].

So that we can proceed, we require the form of the parity-violating nucleon-nucleon potential $V_{NN}^{(-)}$ discussed above. Utilization of this potential, however, requires evaluation of the weak parity violating vertices

$$\langle NM|H_w^{(-)}|N\rangle,$$

where M is one of the mesons that are exchanged—$\pi^{\pm}, \rho, \omega, \ldots$. The obvious way to calculate such matrix elements is to utilize a quark model analysis. Within the valence quark model, we can analyze the ways in which a baryon can fragment into another baryon and an accompanying meson via the weak interaction. Three such diagrams (shown in Figure 5.19) are expected to dominate.

Of these, only the third is directly calculable, being the well-known "factorization" diagram [79] wherein the vector meson is connected to the vacuum through the polar vector current,

$$\langle \rho N|A_\mu V^\mu|N\rangle \simeq \langle \rho|V^\mu|0\rangle\langle N|A_\mu|N\rangle. \tag{5.103}$$

(More precisely, we must also include possible contributions from crossed diagrams that can be obtained from the direct by means of a Fierz transformation [80].) Calculation of the first two diagrams is beyond our ability, as these amplitudes involve complex radial integrals over a priori unknown

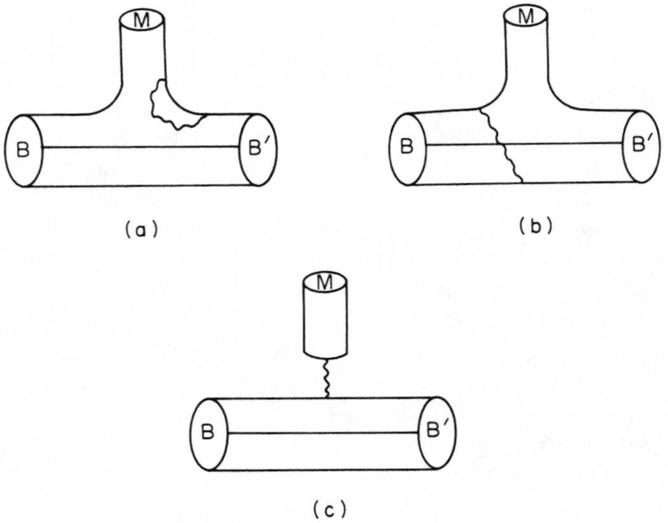

FIGURE 5.19. The three quark topologies that are relevant for nuclear parity violation are depicted here. Diagram (c), the familiar "factorization" diagram, is exactly calculable, while (a) and (b) can be related to corresponding hyperon decay graphs [78].

quark wave functions. However, we *can* make progress by noting that if we are willing to use $SU(6)_w$ symmetry [81], then, since Λ, N, Σ are members of a 56-dimensional representation, while ρ, ω, π are "siblings" within a 35-dimensional representation, the desired

$$\langle \rho N | H_w^{\text{p.v.}} | N \rangle, \qquad \langle \omega N | H_w^{\text{p.v.}} | N \rangle, \qquad \langle \pi N | H_w^{\text{p.v.}} | N \rangle$$

amplitudes can be expressed in terms of *experimentally measured* parity-violating hyperon decay amplitudes, such as

$$\langle \pi N | H_w^{\text{p.v.}} | \Lambda \rangle, \qquad \langle \pi N | H_w^{\text{p.v.}} | \Sigma \rangle.$$

This approach was first advocated by McKellar and Pick [82], but unfortunately, the technique can be applied only for the charged current component of H_w. There, of course, exists a neutral current component, and this piece of the Hamiltonian, mediated by the Z-boson, cannot be directly related to $H_w(\Delta S = 1)$ using $SU(6)_w$ symmetry. Thus, new techniques are needed.

An approach to these amplitudes can be found by noting that one can construct a nucleon from the vacuum using three quark-creation operators,

$$|N\rangle = \underbrace{b^\dagger_{qs} b^\dagger_{q's'} b^\dagger_{q''s''}}_{J=\frac{1}{2},\ I=\frac{1}{2},\ \text{color singlet,}} |0\rangle \tag{5.104}$$

where we imagine the spins, isospins to be combined in such a fashion to form components of a spin, isospin doublet and the colors to be contracted to form a singlet. Similarly, we can construct the vector and pseudoscalar mesons via

$$|M\rangle = \underbrace{b^\dagger_{qs} d^\dagger_{q's'}}_{J,I,\ \text{color singlet,}} |0\rangle \tag{5.105}$$

using a quark and antiquark creation operator. (Here the negative parity arises because the intrinsic parity of fermions and antifermions is opposite, cf. eq. [2.69].) The weak Hamiltonian is itself of current-current form and hence can be written in terms of four quark fields:

$$H_w \sim \frac{G}{\sqrt{2}} \bar{\psi} 0 \psi \bar{\psi} 0' \psi. \tag{5.106}$$

The weak matrix element then has the basic form,

$$\begin{aligned} \langle MN|H_w|N\rangle = {} & \frac{G}{\sqrt{2}} \langle 0|(b_{qs} b_{q's'} b_{q''s''})(b_{qs} d_{q's'}) \\ & \times \bar{\psi} 0 \psi \bar{\psi} 0' \psi (b^\dagger_{qs} b^\dagger_{q's'} b^\dagger_{q''s''})|0\rangle \times R, \end{aligned} \tag{5.107}$$

where R is some complicated radial integral. The vacuum expectation value here is tedious to calculate, but doable. Thus, one finds

$$\langle MN|H_w|N\rangle \sim \text{known "geometrical" factor} \times R, \tag{5.108}$$

which is of a form much like the Wigner-Eckart theorem, where we would call the known "geometrical factor" a Clebsch-Gordan coefficient, and the amplitude R a reduced matrix element, which is the same for all transitions. The constant R may be evaluated empirically by comparing one such amplitude with its experimental value. In fact, when this procedure is followed for the simple charged current Hamiltonian, the $SU(6)_w$ predictions are exactly reproduced, and the "geometrical factor" is indeed just an SU(6) Clebsch-Gordan coefficient. However, the quark model technique can also be extended straightforwardly to the neutral current Hamiltonian, and this is the procedure by which such amplitudes are evaluated.

Several subleties, however, are worthy of notice. One is that although obtaining the π-emission amplitudes really only requires SU(3) symmetry, the calculation of the vector meson transitions, of necessity, utilizes the larger group SU(6). Unfortunately, SU(6) is rather badly broken for mesons since, e.g., although π and ρ mesons are members of a common 35-dimensional representation and should therefore be degenerate in the exact SU(6) limit, experimentally we have

$$m_\pi \sim 140 \text{ MeV}, \qquad m_\rho \sim 760 \text{ MeV}.$$

A believable calculation then requires techniques that include some of these symmetry-breaking effects.

A second important feature of the calculation is that by renormalization group techniques, hard gluon corrections to the basic weak Hamiltonian are included. For example, with the effective weak Hamiltonian written as

$$\begin{aligned} \mathscr{H}_w = &\frac{G}{(2)^{1/2}} M_W{}^2 \int d^4x D_F(x, M_W) T(J_\mu{}^+(x) J^{-\mu}(0)) \\ &+ \frac{8G}{(2)^{1/2}} M_Z{}^2 \int d^4x D_F(x, M_Z) T(J_\mu{}^Z(x) J^{Z\mu}(0)) \end{aligned} \tag{5.109}$$

the $\Delta S = 1$ renormalized form is given by [83].

$$\mathscr{H}_w(\Delta S = 1) = \frac{2G}{2^{1/2}} \cos\theta_c \sin_c[K^{0.48}\mathscr{O}_- + K^{-0.24}\mathscr{O}_+] + \text{h.c.}, \tag{5.110}$$

with

$$K = 1 + \frac{g^2(\mu^2)}{16\pi^2} b \ln \frac{M_W{}^2}{\mu^2} \tag{5.111}$$

being the usual renormalization group factor, and

$$\begin{aligned} \mathscr{O}_- &= \frac{1}{4}\{\bar{d}\gamma_\mu(1+\gamma_5)u\bar{u}\gamma^\mu(1+\gamma_5)s - \bar{d}\gamma^\mu(1+\gamma_5)s\bar{u}\gamma_\mu(1+\gamma_5)u\}, \\ \mathscr{O}_+ &= \frac{1}{4}\{\bar{d}\gamma_\mu(1+\gamma_5)u\bar{u}\gamma^\mu(1+\gamma_5)s + \bar{d}\gamma^\mu(1+\gamma_5)s\bar{u}\gamma_\mu(1+\gamma_5)u\}. \end{aligned} \tag{5.112}$$

Reasonable values for K are

$$3 < K < 5, \tag{5.113}$$

depending on specific assumptions about the renormalization point μ.

The $\Delta S = 1$ Hamiltonian above involves only charged currents and is relatively simple. On the other hand, its $\Delta S = 0$ counterpart includes both charged *and* neutral current components and has a more complex, but calculable, form. Defining

$$O(M,N) = \bar{q}\gamma_\mu\gamma_5 Mq\bar{q}\gamma^\mu Nq, \tag{5.114}$$

we have

$$\begin{aligned}\mathscr{H}_w^{\text{eff}}(\Delta S = 0) = \frac{G}{2^{1/2}} \cos\theta_c \sin\theta_c \Bigg[\sum_{i=1}^{2} (\alpha_{ii} O(A_i{}^+, A_i) \\ + \beta_{ii} O(A_i{}^+ t_A, A_i t_A) + \text{h.c.}) \\ + \sum_{i,j=1}^{2} (\gamma_{ij} O(B_i, B_j) + \delta_{ij} O(B_i t^A, B_j t^A)) \Bigg],\end{aligned} \tag{5.115}$$

where t^A are SU(3) color matrices normalized via $Trt^A t^B = 2\delta^{AB}$,

$$\begin{aligned} q &= \begin{pmatrix} u \\ d \\ s \end{pmatrix} \\ A_1 &= \begin{pmatrix} 0 & 1 & 0 \\ 0 & 0 & 0 \\ 0 & 0 & 0 \end{pmatrix}, \qquad A_2 = \begin{pmatrix} 0 & 0 & 1 \\ 0 & 0 & 0 \\ 0 & 0 & 0 \end{pmatrix} \\ B_1 &= \begin{pmatrix} 1 & 0 & 0 \\ 0 & 1 & 0 \\ 0 & 0 & 1 \end{pmatrix}, \qquad B_1 = \begin{pmatrix} 1 & 0 & 0 \\ 0 & -1 & 0 \\ 0 & 0 & -1 \end{pmatrix} \end{aligned} \tag{5.116}$$

and

$$\begin{aligned} \alpha_{11} &= \cot\theta_c (K^{0.48} + 2K^{-0.24})/3, \\ \alpha_{22} &= \tan^2\theta_c \alpha_{11}, \\ \beta_{11} &= \cot\theta_c (-K^{0.48} + K^{-0.24})/4, \\ \beta_{22} &= \tan^2\theta_c \beta_{11}, \end{aligned}$$

$$
\begin{aligned}
\frac{1}{2}\sin 2\theta_c\gamma_{11} &= (3 - 2\sin^2\theta_w)[0.0556K^{0.48} \\
&\qquad - 0.051K^{0.35} - 0.067K^{-0.24} + 0.062K^{-0.40}], \\
\frac{1}{2}\sin 2\theta_c\gamma_{12} &= -\frac{1}{3}\sin^2\theta_w[-0.049K^{0.85} \\
&\qquad + 0.190K^{0.43} - 0.426K^{-0.13} + 0.27K^{-0.35}], \\
\frac{1}{2}\sin 2\theta_c\gamma_{21} &= -\frac{1}{3}\sin^2\theta_w[0.086K^{0.85} \\
&\qquad + 0.146K^{0.43} + 0.623K^{-0.13} + 0.151K^{-0.35}], \\
\frac{1}{2}\sin 2\theta_c\gamma_{22} &= (1 - 2\sin^2\theta_w)[0.167K^{0.48} + 0.333K^{-0.24}], \\
\frac{1}{2}\sin 2\theta_c\delta_{11} &= (3 - 2\sin^2\theta_w)[-0.042K^{0.48} \\
&\qquad + 0.028K^{0.35} - 0.025K^{-0.13} + 0.039K^{-0.40}], \\
\frac{1}{2}\sin 2\theta_c\delta_{12} &= -\frac{1}{3}\sin^2\theta_w[-0.113K^{0.85} \\
&\qquad - 0.099K^{0.43} + 0.129K^{-0.13} + 0.079K^{-0.35}], \\
\frac{1}{2}\sin 2\theta_c\delta_{21} &= -\frac{1}{3}\sin^2\theta_w[0.063K^{0.85} \\
&\qquad - 0.126K^{0.43} - 0.084K^{-0.13} + 0.148K^{-0.35}], \\
\frac{1}{2}\sin 2\theta_c\delta_{22} &= (1 - 2\sin^2\theta_w)(-K^{0.48} + K^{-0.24})/8.
\end{aligned}
\tag{5.117}
$$

(Of course, the Weinberg angle enters into these results for the neutral current piece.)

Using this effective weak Hamiltonian and the quark model techniques outlined earlier, we can now evaluate the parity violating couplings, which can be parameterized as

$$
\begin{aligned}
\mathscr{H}^{\text{p.v.}}_{MNN} &= (2)^{-1/2} f_\pi \bar{N}(\vec{\tau} \times \vec{\phi}^\pi)^3 N \\
&\quad + \bar{N}\left[h_\rho{}^0 \vec{\tau} \cdot \vec{\phi}_\mu{}^\rho + h_\rho{}^1 \phi_\mu^{\rho 3} + h_\rho{}^2 \frac{(3\tau^3\phi_\mu^{\rho 3} - \vec{\tau} \cdot \vec{\phi}_\mu{}^\rho)}{2(6)^{1/2}}\right]\gamma^\mu\gamma_5 N \\
&\quad + \bar{N}[h_\omega{}^0 \phi_\mu{}^\omega + h_\omega{}^1 \tau^3 \phi_\mu{}^\omega]\gamma^\mu\gamma_5 N - h_\rho'^1 \bar{N}(\vec{\tau} \times \vec{\phi}_\mu{}^\rho)^3 \frac{\sigma^{\mu\nu}k_\nu}{2M}\gamma_5 N.
\end{aligned}
\tag{5.118}
$$

Here, h_v^{I} refers to the $N \to NV$ coupling carrying isospin I. (The pion coupling is purely isovector and is conventionally written as f_π.) These weak amplitudes are then combined with their strong interaction counterparts,

$$\mathscr{H}^{\text{p.c.}}_{MNN} = ig_{\pi NN}\bar{N}\gamma_5\vec{\tau}\cdot\vec{\phi}^{\pi}N + g_\rho\bar{N}\left(\gamma_\mu + \frac{i\chi_v}{2M}\sigma_{\mu\nu}k^\nu\right)\vec{\tau}\cdot\vec{\phi}^{\mu\rho}N$$
$$+ g_\omega\bar{N}\left(\gamma^\mu + \frac{i\chi_s}{2M}\sigma^{\mu\nu}k_\nu\right)\phi_\mu{}^\omega N, \tag{5.119}$$

in order to generate an effective parity-violating nucleon-nucleon potential, which has the form

$$V^{\text{p.v.}}_{12} = \frac{f_\pi g_{\pi NN}}{2^{1/2}}\, i\left(\frac{\vec{\tau}_1\times\vec{\tau}_2}{2}\right)^3(\vec{\sigma}_1+\vec{\sigma}_2)\cdot\left[\frac{\vec{p}_1-\vec{p}_2}{2M}, f_\pi(r)\right]$$
$$- g_\rho(h_\rho{}^0\vec{\tau}_1\cdot\vec{\tau}_2 + h_\rho{}^1\left(\frac{\vec{\tau}_1+\vec{\tau}_2}{2}\right)^3 + h_\rho{}^2\frac{(3\tau_1{}^3\tau_2{}^3-\vec{\tau}_1\cdot\vec{\tau}_2)}{2(6)^{1/2}}$$
$$\times\left((\vec{\sigma}_1-\vec{\sigma}_2)\cdot\left\{\frac{\vec{p}_1-\vec{p}_2}{2M}, f_\rho(r)\right\}\right.$$
$$\left.+ i(1+\chi_v)\vec{\sigma}_1\times\vec{\sigma}_2\cdot\left[\frac{\vec{p}_1-\vec{p}_2}{2M}, f_\rho(r)\right]\right.$$
$$- g_\omega\left(h_\omega{}^0 + h_\omega{}^1\left(\frac{\vec{\tau}_1+\vec{\tau}_2}{2}\right)^3\right) \tag{5.120}$$
$$\times\left((\vec{\sigma}_1-\vec{\sigma}_2)\cdot\left\{\frac{\vec{p}_1-\vec{p}_2}{2M}, f_\omega(r)\right\}\right.$$
$$\left.+ i(1+\chi_s)\vec{\sigma}_1\times\vec{\sigma}_2\cdot\left[\frac{\vec{p}_1-\vec{p}_2}{2M}, f_\omega(r)\right]\right)$$
$$- (g_\omega h_\omega{}^1 - g_\rho h_\rho{}^1)\left(\frac{\vec{\tau}_1-\vec{\tau}_2}{2}\right)^3(\vec{\sigma}_1+\vec{\sigma}_2)\cdot\left\{\frac{\vec{p}_1-\vec{p}_2}{2M}, f_\rho(r)\right\}$$
$$- g_\rho h_\rho'^1 i\left(\frac{\vec{\tau}_1\times\vec{\tau}_2}{2}\right)^Z(\vec{\sigma}_1+\vec{\sigma}_2)\cdot\left[\frac{\vec{p}_1-\vec{p}_2}{2M}, f_\rho(r)\right].$$

where $f_v(r) = \exp(-m_v r)/4\pi r$ is the usual Yukawa potential.

Rather than quote precise values for these weak couplings, it is prudent, in view of the large uncertainties associated with symmetry-breaking and strong-interaction effects, to give an allowed range together with a "best" guess for what the number would be if these ambiguities were under control. Results are quoted in Table 5.3 for the Cabibbo (charged account) piece and in Table 5.4 for the full Weinberg-Salam form. Note that the

TABLE 5.3
CALCULATED VALUES FOR *NNM* WEAK COUPLINGS IN THE CABIBBO (Charged Current) MODEL

Cabibbo Model	Reasonable Range	"Best" Value
f_π	$0 \rightarrow 1$	0.5
$h_\rho^{\ 0}$	$15 \rightarrow -64$	-25
$h_\rho^{\ 1}$	$0 \rightarrow -0.7$	-0.4
$h_\rho^{\ 2}$	-58	-58
$h_\omega^{\ 0}$	$6 \rightarrow -22$	-6
$h_\omega^{\ 1}$	$0 \rightarrow -2$	-1
$h_\phi^{\ 0}$	0.2	0.2
$h_\phi^{\ 1}$	0.3	0.3

Shown is the overall range as well as the "most likely" value as given by DDH [78] for weak parity-violating *NNM* vertices, as calculated for the charged current (Cabibbo) weak interaction. Amplitudes are in units of $g_w = 3.8 \times 10^{-8}$.

TABLE 5.4
CALCULATED VALUES FOR *NNM* WEAK COUPLINGS IN THE WEINBERG-SALAM (Charged Plus Neutral Current) MODEL

Weinberg-Salam Model	Reasonable Range	"Best" Value
f_π	$0 \rightarrow 30$	12
$h_\rho^{\ 0}$	$30 \rightarrow -81$	-30
$h_\rho^{\ 1}$	$-1 \rightarrow 0$	-0.5
$h_\rho^{\ 2}$	$-20 \rightarrow -29$	-25
$h_\omega^{\ 0}$	$15 \rightarrow -27$	-5
$h_\omega^{\ 1}$	$-5 \rightarrow -2$	-3
$h_\phi^{\ 0}$	$0.4 \rightarrow 0.2$	0.3
$h_\phi^{\ 1}$	$-20 \rightarrow -13$	-13

Shown is the overall range as well as the "most likely" value as given by DDH [78] for weak parity-violating *NNM* vertices, as calculated for the charged plus neutral current (Weinberg-Salam) weak interaction. Amplitudes are in units of $g_w = 3.8 \times 10^{-8}$.

simple charged current and full Weinberg-Salam couplings are basically similar except for the case of f_π, which is suppressed by $\sin^2\theta < \sim 0.04$ in the Cabibbo case. This is because the leading ($\cos^2\theta_c$) contributions result from the *symmetric* product of $I = 1$ charged currents and therefore appear only in the $\Delta I = 0,2$ channels.

It remains to evaluate the nuclear matrix elements of $V_{NN}^{(-)}$ in order to touch base with the experimental observables. For the longitudinal asymmetry in light systems this is relatively straightforward, yielding [84], e.g.,

$$\begin{aligned} \vec{p}p\,(45\text{ MeV}) \quad & A_L \simeq 0.175\left(0.32g_\rho\left(h_\rho{}^0 + \frac{1}{\sqrt{6}}h_\rho{}^2\right) + 0.10g_\omega h_\omega{}^0\right) \\ \vec{p}\alpha\,(45\text{ MeV}) \quad & A_L \simeq -0.026g_{\pi NN}f_\pi + 0.050g_\rho h_\rho{}^0 + 0.007g_\omega h_\omega{}^0. \end{aligned} \tag{5.121}$$

With only two pieces of data and four independent weak amplitudes, we cannot go much further without additional assumptions. We *can* proceed by using the nucleus as a weak interaction amplifier as outlined above. However, we pay a price for this increased signal by having correspondingly larger "noise" introduced by nuclear wave function uncertainties. Thus, for example, one can utilize the shell model in order to express the experimental observables in ^{19}F and ^{21}Ne in terms of the weak couplings [84],

$$\begin{aligned} {}^{19}\text{F} \quad & A_\gamma = -248f_\pi + 105(h_\rho{}^0 + 0.56h_\omega{}^0) \\ {}^{21}\text{Ne} \quad & P_\gamma = 90260f_\pi + 30790(h_\rho{}^0 + 0.56h_\omega{}^0). \end{aligned} \tag{5.122}$$

Note that the coefficients of the weak couplings are much larger than in the case of $\vec{p}p$ and $\vec{p}\alpha$ scattering quoted above. This shows dramatically the amplification factor present in these nuclear systems. In principle, we could now proceed with our analysis using these experimental numbers. However, before doing so, it is necessary to look at this calculation more carefully. The nuclear matrix elements have been calculated in a simple $0\hbar\omega$, $1\hbar\omega$ shell model basis. However, any realistic state also contains an additional $2\hbar\omega$ component, which is generally small ($\varepsilon \ll 1$) if the simple shell model picture is to be valid:

$$|\psi(+)\rangle = |0\hbar\omega\rangle + \varepsilon|2\hbar\omega\rangle. \tag{5.123}$$

If we use the simple shell model to calculate observables such as the Gamow-Teller matrix element $\langle|\sum_i \tau_i\boldsymbol{\sigma}_i|\rangle$ or another similar operator that does not connect $|0\hbar\omega\rangle$ and $|2\hbar\omega\rangle$ levels, then a fairly reliable estimate should result, since corrections are $\mathcal{O}(\varepsilon^2)$:

$$\langle\psi_+|\mathcal{O}|\psi_+\rangle = \langle\psi_+^{(0)}|\mathcal{O}|\psi_+^{(0)}\rangle + \mathcal{O}(\varepsilon^2). \tag{5.124}$$

However, for the parity mixing term we are coupling to an odd parity $1\hbar\omega$ level that can connect either to $0\hbar\omega$ or $2\hbar\omega$ levels. Thus corrections to the simple shell model calculation are $\mathcal{O}(\varepsilon)$:

$$\langle\psi_-|H_w^{\text{p.v.}}|\psi_+\rangle = \langle\psi_-^{(0)}|H_w^{\text{p.v.}}|\psi_+^{(0)}\rangle + \mathcal{O}(\varepsilon). \tag{5.125}$$

In fact, by doing a large $0\hbar\omega$, $1\hbar\omega$, $2\hbar\omega$ basis calculation, Haxton has found in the case of ^{18}F that [85]

$$\frac{\langle\psi_-|H_w^{\text{p.v.}}|\psi_+\rangle_{0,1,2\hbar\omega}}{\langle\psi_-|H_w^{\text{p.v.}}|\psi_+\rangle_{0,1\hbar\omega}} \simeq \frac{1}{3}, \tag{5.126}$$

so that these so-called core polarization effects are not small. The trouble is that they are also model dependent and do not offer much hope for a reliable estimate.

There does exist a way, however, in which to "measure" this effect experimentally. It has been pointed out that the form of the parity violating nucleon-nucleon potential relevant for π-exchange [86],

$$V_{NN}^{(-)}(\pi \text{ exch}) = \frac{i}{2\sqrt{2}} f_\pi g_{\pi NN}(\tau_1 \times \tau_2)_3(\boldsymbol{\sigma}_1 + \boldsymbol{\sigma}_2)\cdot\left[\frac{\mathbf{p}_1 - \mathbf{p}_2}{2m_N}, \frac{e^{-m_\pi r}}{r}\right], \tag{5.127}$$

is an isotopic partner of the two-body pion exchange contribution to the timelike piece of the axial current A_0, which is probed in nuclear beta decay:

$$\begin{aligned} A_0 = A_0(\text{1-body}) &+ \frac{i}{2} g_A g_{\pi NN}(\tau_1 \times \tau_2)_\pm(\boldsymbol{\sigma}_1 + \boldsymbol{\sigma}_2) \\ &\cdot\left[\frac{\mathbf{p}_1 - \mathbf{p}_2}{2m_N}, \frac{e^{-m_\pi r}}{r}\right]. \end{aligned} \tag{5.128}$$

If somehow this two-body matrix element of A_0 could be measured in a beta-decay transition between levels isotopically related to those involved in the weak parity mixing, then the weak pion-exchange matrix element can be determined experimentally! The problem with this idea is that there is no way to separate the two-body and one-body contributions—the matrix element found experimentally is the sum of the two. Nevertheless, Haxton has noted that the ratio of two-body and one-body terms [87],

$$\frac{\langle|A_0(\text{2-body})|\rangle}{\langle|A_0(\text{1-body})|\rangle} \simeq 0.6 \tag{5.129}$$

is to a large extent model *independent*. Using this assumption, we can then "measure" $\langle|A_0(\text{2-body})|\rangle$, and when this is done in the case of ^{19}F, we find that [88]

$$\frac{\left\langle \frac{1}{2}^- \middle| V_{NN}^{(-)} \middle| \frac{1}{2}^+ \right\rangle^{\text{exp}}}{\left\langle \frac{1}{2}^- \middle| V_{NN}^{(-)} \middle| \frac{1}{2}^+ \right\rangle_{0,1\hbar\omega}} = 0.34 \pm 0.03, \tag{5.130}$$

in good agreement with the value calculated above using the large $0\hbar\omega$, $1\hbar\omega$, $2\hbar\omega$ shell model basis. Unfortunately, one does *not* have a similar method to "measure" ρ and ω matrix elements. However, there exists reason to believe that such matrix elements may scale similarly to that for pion exchange [87].

We can finally proceed with our fit to the $\vec{p}p$, $\vec{p}\alpha$, ^{19}F, and ^{21}Ne data, which are represented in Figure 5.20 in terms of bands in f_π, h_ρ^0 space. We observe that these experiments are *not* independent. In fact results for ^{19}F, $\vec{p}\alpha$, and $\vec{p}p$ are quite consistent with one another and basically correspond to the same band in parameter space, although ^{19}F, because of the better statistics of the experiment, has the smaller error bar. The ^{21}Ne experiment, however, represents a very different region in this space. (The reason for this is quite clear—^{19}F, $\vec{p}\alpha$, $\vec{p}p$ involve an unpaired *proton* while ^{21}Ne

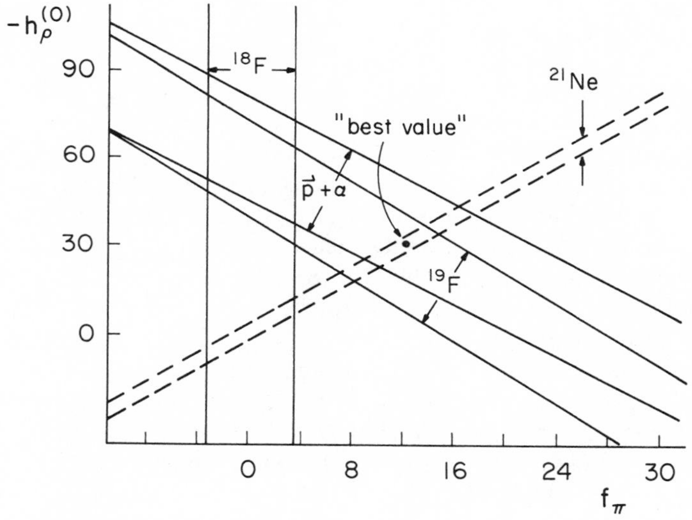

FIGURE 5.20. Plotted are the ranges in h_ρ^0, f_π space allowed by various nuclear parity violation experiments. Note that results from $\vec{p} + \alpha$, ^{19}F, and ^{21}Ne experiments are consistent with the "best value" theoretical estimates. However, the very precise ^{18}F measurements require a much smaller value of f_π. (E. Adelberger, in AIP Conf. Proc. No. 150 (1986) p. 1177).

involves an unpaired *neutron* state.) The intersection region of the two bands includes the best-value estimates of the theoretical calculations, and thus we might be tempted to conclude that our understanding of nuclear parity violation is secure. Unfortunately, this is not the case. The problem lies with the ^{18}F experiments, for which theoretical analysis yields [89]

$$|P_\gamma| = 4340 f_\pi \tag{5.131}$$

Here the dependence solely on f_π occurs since the parity mixing is purely $\Delta I = 1$. The coefficient in front of the pion coupling f_π is thought to be fairly reliable since it is normalized via measurement of the analog beta decay,

$$^{18}\text{Ne} \to {}^{18}\text{F}(0^-, 1.081 \text{ MeV}) + e^+ + \nu_e,$$

as described above for the ^{19}F system [90]. Comparison with the experimental bound

$$|P_\gamma| < 5 \times 10^{-4} \tag{5.132}$$

then yields the vertical band shown in Figure 5.20, which is obviously inconsistent with the intersection region found earlier. In fact, the ^{18}F experiments yield an upper limit on f_π

$$f_\pi < 3.8 g_w, \tag{5.133}$$

which is considerably smaller than the best-value number given in Table 5.4.

It is thought that the inconsistency here most likely lies with the analysis of the ^{21}Ne experiment. One has no confirming "odd neutron" experimental evidence, and the theoretical calculation of this system—involving five nucleons above a closed ^{16}O shell—is notoriously difficult. Reliable large basis calculations for ^{21}Ne will require a new generation of computational power. If this is indeed the source of the problem, then the upper bound on the size of the weak parity violating pion coupling given in eq. (5.133) presents an interesting challenge to theorists and may provide useful insights into the reliability of such quark model calculations. The point is that the estimated value of f_π is the sum of three separate components, each of which is thought to have the same sign [91]. Thus each piece must obey the bound in eq. (5.133) and this could prove a problem if the experimental number is pushed any lower.

So what does the future hold? On the theoretical side it is important to (1) better understand and update the quark model calculations of the weak parity violating couplings and in addition; and (2) check the validity of the associated nuclear wavefunction calculations which make the connection with experimental observables, especially for the ^{21}Ne system. Experimentally there is ongoing work in a number of different areas [92]. One of the most interesting measurements is that being performed by a University of Washington/Wisconsin collaboration on mixing of the 0^+0^- doublet in the $A = 14$ system, which is primarily sensitive to $h_\rho{}^0$ [93]. At the present time there appears to exist a possible discrepancy between the preliminary experimental result [93]

$$\langle H_{wk} \rangle = +0.36 \pm 0.35 \text{ eV} \tag{5.134a}$$

and a simple theoretical calculation [94]

$$\langle H_{wk} \rangle = -1.04 \text{ eV}. \tag{5.134b}$$

A second area of activity is an approved TRIUMF experiment that will measure the longitudinal asymmetry of polarized p–p scattering at 230 MeV [95]. Such a measurement will provide an independent probe of $h_\rho{}^0$ in a system which does not involve nuclear complications. Finally, a U.S./French/German collaboration proposes to continue their work on neutron spin rotation at Grenoble, extending their measurements from heavy nuclei such as ^{139}La, where effects have already been seen [96] but where reliable calculations are impossible, to light nuclei, wherein effects are smaller but calculable [97].

Clearly, the immediate goal in this field must be to verify the validity of the simple meson exchange picture—for if this approach itself is inconsistent, we must begin again at ground zero. In any case, despite the best efforts of experimentalists and theorists alike, it appears that the subject of nuclear parity violation will remain an active area of physics for some time to come.

5.3 Hypernuclear Decay

I shall now discuss a topic which does not really belong in the symmetry category—hypernuclear weak decay. However, because of its similarity to the problem of nuclear parity violation, I have chosen to discuss it at this point. This process is one that is just beginning to be studied and will no doubt be a focus of future theoretical and experimental work.

First, though, let's go back a step and ask what *is* a hypernucleus. Compared to a conventional nuclear system made up of Z protons and A–Z neutrons, a Λ-hypernucleus has Z protons, A–Z–1 neutrons, and a lambda hyperon. At some level, one might think that the Λ, having charge and mass similar to the neutron, should not have much effect. However, this

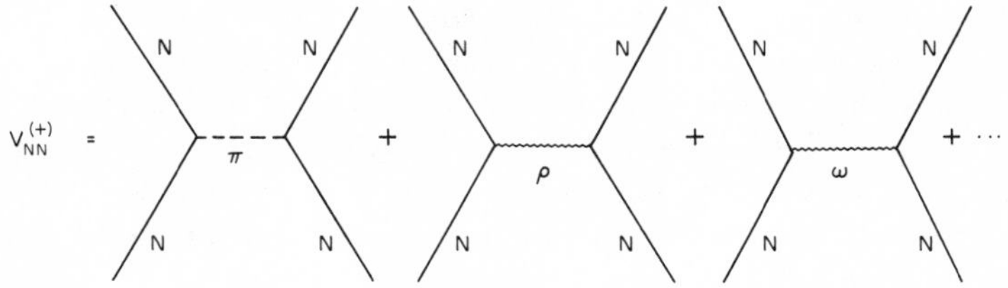

FIGURE 5.21. The ordinary strong nucleon-nucleon potential is reasonably well represented in terms of the sum of various single meson exchange diagrams—π, ρ, ω, σ, etc.

is not at all the case, because the lambda differs from the neutron also in its isospin and strangeness quantum numbers:

$$\Lambda: \begin{array}{l} I = 0 \\ S = -1 \end{array} \qquad n: \begin{array}{l} I = \frac{1}{2} \\ S = 0 \end{array}$$

The importance of this isospin difference to the $\Lambda - N$ interaction is understood most easily in a meson exchange model where we envision the NN potential as arising from the exchange of π, η, ρ, ω, σ . . . mesons between the nucleon pair (cf. Figure 5.21). However, since the Λ is an isoscalar, only $I = 0$ mesons such as the ω are permitted in the single boson exchange picture (cf. Figure 5.22); single ρ- and π-exchange are forbidden. Consequently, the $\Lambda - N$ system is expected to have a very different potential than from its $N - N$ counterpart [98].

The difference in strangeness quantum number between Λ and n also plays an important role. In free space, of course, both Λ, n are unstable,

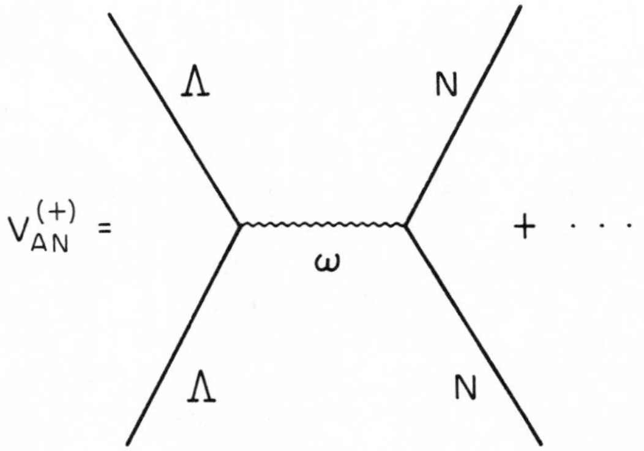

FIGURE 5.22. The strong ΛN potential does not involve isovector—π, ρ, . . .—exchange and hence has a completely different character than its NN counterpart.

with

$$n \to p + e^- + \bar{\nu}_e \qquad \tau_n \sim 9 \times 10^2 \text{ sec}$$
$$\Lambda \to \begin{matrix} p + \pi^- \\ n + \pi^0 \end{matrix} \qquad \tau_\Lambda \sim 3 \times 10^{-10} \text{ sec.} \tag{5.134}$$

However, since

$$m_n - m_p = 1.3 \text{ MeV}, \tag{5.135}$$

while nuclear binding energies are of order 8 MeV/nucleon a bound neutron is often effectively stable; its decay is kinematically forbidden. On the other hand, since

$$m_\Lambda - (m_N + m_\pi) \sim 40 \text{ MeV} \gg B_\Lambda, \tag{5.136}$$

a Λ confined within a hypernucleus, remains unstable and eventually decays, which is the topic of our study.

First let's ask how hypernuclei are produced. Often a (K^-,π^-) reaction is used

$${}^A N_Z(K^-,\pi^-){}^A_\Lambda N_Z$$

The ideal momentum of the K-beam in the lab is in the vicinity of 600 MeV/c, since at this point, as shown in Figure 5.23, the kinematics are

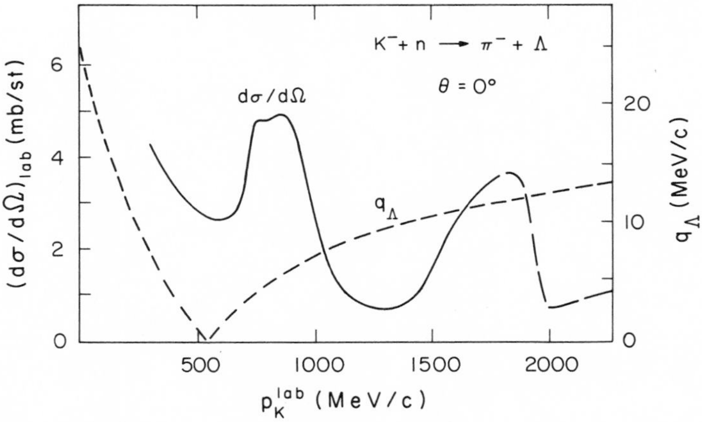

FIGURE 5.23. Shown is the experimental cross section for forward Λ production in $K^-n \to \pi^-\Lambda$ as well as the recoil Λ momentum, both as a function of lab momentum. Note that in the vicinity of $p_K^{\text{lab}} \approx 500$ MeV/c, the Λ is deposited essentially at rest. (P. D. Barnes, Workshop on Nuclear and Particle Physics, LANL, Los Alamos, NM (1981)).

such that the Λ is produced essentially at rest in the hypernuclear system. (Actually the energy used is usually somewhat higher—800 MeV/c—in order to take advantage of the increased cross section in this region.) If indeed the Λ is deposited with little momentum transfer, then the Λ finds itself captured into a ${}^{J}L_{1/2}$ shell-model state—in the Λ-shell model basis.

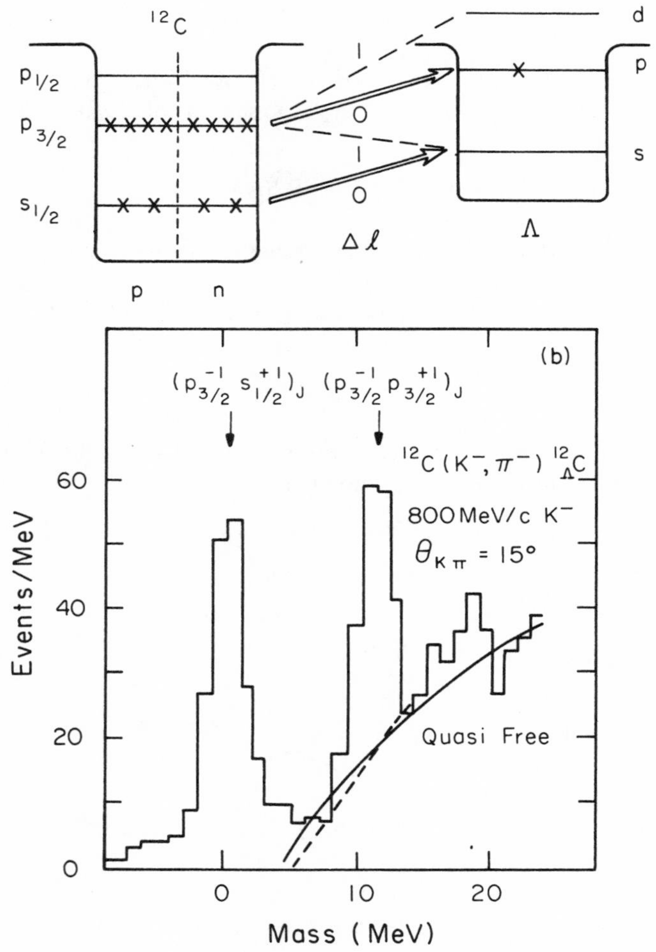

FIGURE 5.24. Hypernuclear production can be thought of as replacing a neutron from some shell model orbital by a Λ in the same or different Λ-shell model state. Both examples are seen in the experimental data shown. (P. D. Barnes, Workshop on Nuclear and Particle Physics, LANL, Los Alamos, NM (1981)).

That is, there is an effective mean field potential made up of the sum of ΛN interactions, and it is in *this* "shell model" basis that the Λ finds itself. However, all states but one in this space are empty. The lambda will typically find itself in a low-lying state, and study of the spectroscopy of such states has been used to give insight into the ΛN interaction. Thus, in Figure 5.24 we see that in a (K^-,π^-) reaction on ^{12}C, which produces the hypernucleus $^{12}_{\Lambda}$C, a neutron is removed from the $1P_{3/2}$ shell and reincarnated as a lambda in either a $1P_{3/2}$ or $1S_{1/2}$ lambda-shell model state. Study of the splittings between such levels and between states split by $\Lambda - N$ fine structure has revealed a lambda-nucleon potential which has very little spin-orbit interaction and whose central component is somewhat weaker than its nucleon-nucleon counterpart [99]. Both of these results are consistent with the boson exchange considerations discussed above.

However, analysis of the $\Lambda - N$ interaction is not our purpose here. Rather, we note that regardless of which shell model state the lambda is captured in, it will quickly cascade down to the lowest $(1S_{1/2})$ level, emitting gamma rays in the process. But then what? A first thought might well be that the lambda decays essentially as in free space,

$$\Lambda \begin{array}{l} \nearrow p\pi^- \\ \searrow n\pi^0 \end{array},$$

but with a rate modified slightly by the nuclear medium. That this is not the case, however, can be seen by calculating the energy and momentum of the nucleon emitted in such a transition for the case of a free Λ:

$$\begin{aligned} KE_N &\simeq \frac{(m_\Lambda - m_N)^2 - m_\pi{}^2}{2m_\Lambda} \sim 5 \text{ MeV} \\ p_N &= \sqrt{2m_N KE_N} \sim 100 \text{ MeV/c}. \end{aligned} \tag{5.137}$$

Thus, the momentum is below the Fermi level for most nuclei, and the $N\pi$ decay is Pauli blocked. A theoretical calculation of this effect is shown in Figure 5.25, where we can see that the mesonic decay rate of a hypernucleus is suppressed by two orders of magnitude already by $A \approx 15$. Thus the usual mesonic decay channels are effectively shut out. This would seem to lead to hypernuclear lifetimes which are extremely long compared to the 260 picosecond lifetime of a free lambda. However, Cheston and Primakoff pointed out long ago that an alternative decay mechanism is available to a lambda within the nuclear medium [100]. The Λ can decay

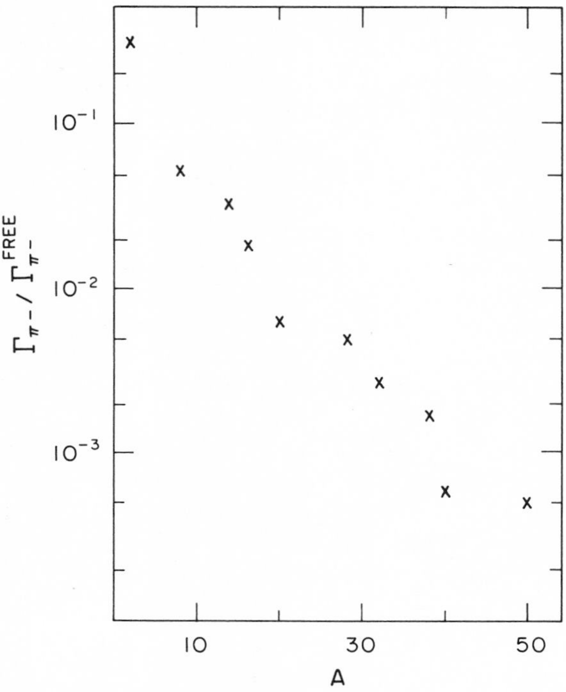

FIGURE 5.25. Plotted are results of a simple theoretical calculation of the Pauli blocking effect on π^- decay of hypernuclei as a function of atomic mass number. The decay rate is well under 10% of that for the corresponding free Λ process once $A \gtrsim 10$.

by interacting with one of its fellow nucleons:

$$\Lambda + N \to N + N.$$

In this situation, assuming that the energy is split evenly between the two outgoing nucleons, we have

$$\begin{aligned} KE_N &\sim \frac{m_\Lambda - m_N}{2} \sim 90 \text{ MeV} \\ p_N &= \sqrt{2m_N KE_N} \sim 400 \text{ MeV/c}, \end{aligned} \tag{5.138}$$

which is well above the Fermi surface, allowing this decay mechanism to proceed essentially uninhibited. That this expectation is indeed borne out

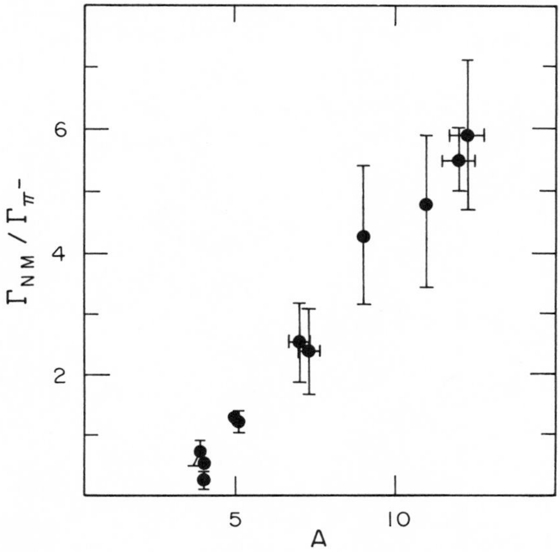

FIGURE 5.26. Plotted is the ratio of nonmesonic to π^- decay rates of hypernuclei as a function of mass number. Note that the nonmesonic mode is dominant for $A \gtrsim 10$. (J. F. Dubach in *Weak and Electromagnetic Interactions in Nuclei*, ed. H. Klapdor, Springer, New York (1986)).

can be seen from the limited experimental data on the process as shown in Figure 5.26, where is plotted the rate Γ_{NM} for the nonmesonic decay $\Lambda N \rightarrow NN$ divided by the corresponding mesonic rate Γ_{π^-} for decay into the π^- channel. Clearly, $\Gamma_{NM} \gg \Gamma_{\pi^-}$ once $A \gtrsim 8$.

What else is interesting to examine in weak hypernuclear decay? One of the obvious observables is the hypernuclear lifetime itself. As noted above, a free Λ decays via the $N\pi$ channel in a time $\tau_\Lambda \sim 260$ pico sec. The Pauli blocking in a nucleus lengthens this lifetime; the opening of the new nomesonic channel leads to a shortening. Which process wins? A second quantity of interest is obtained by noting that the Λ can interact either with a neighboring neutron *or* proton,

$$\Lambda + p \rightarrow n + p$$
$$\Lambda + n \rightarrow n + n,$$

yielding a different final state signal depending on which channel is operative. It is interesting to ask, which mechanism is more important? Before answering these very basic questions, let us briefly review some of the early work in this field.

During the late 1950s and early 1960s, Dalitz and collaborators[104] examined this system theoretically using the assumptions of [101] (1) a local $\Lambda - N$ interaction—$\delta^3(\mathbf{r}_\Lambda - \mathbf{r}_N)$; and (2) incoherence between interactions with differing nucleons,

$$\Gamma_\Lambda^{\text{Tot}} = \sum_i \Gamma_{\Lambda N_i}.$$

These investigations found, e.g.,

$$\begin{aligned} &\Gamma_{NM} \sim \frac{1}{2}\Gamma_\Lambda^{(0)} \qquad {}^5\text{He}_\Lambda \\ &\Gamma_{NM} \sim 2\Gamma_\Lambda^{(0)} \qquad \text{Nuclear Matter.} \end{aligned} \tag{5.139}$$

An improved and more detailed calculation was undertaken in 1965 by J. Barkeley Adams, at the time a post-doc at Stanford [102]. In this model, one includes all possible meson exchanges between ΛN and NN vertices, using one strong and one weak vertex (see Figure 5.27). We have seen very similar diagrams in our earlier discussion of nuclear parity violation. However, there are important differences in the hypernuclear application: (1) there is no Barton's theorem, so π^0, η^0, etc., exchanges must be present; and (2) because this is a strangeness-changing process, K^0, K^{0*}, etc., poles must be included.

Adams's calculation included these features, and in addition utilized short-range correlations for both initial and final state systems. He determined that only pion exchange, presumably because of its long-range

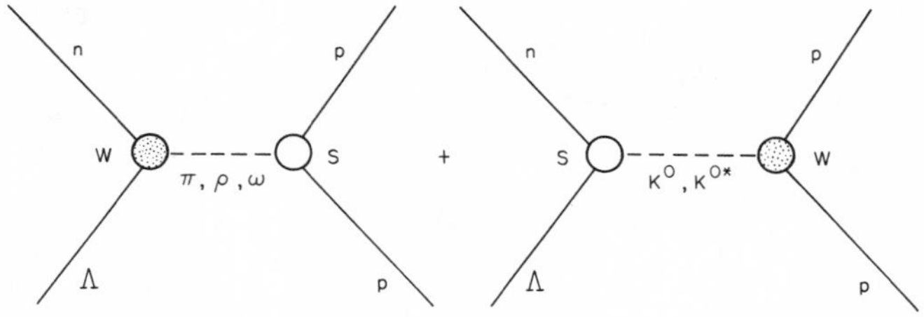

FIGURE 5.27. In the simple meson exchange picture for nonmesonic hypernuclear decay, the exchange of both nonstrange (π, ρ, ω, ...) and strange (K^0, K^{0*}, ...) mesons must be included.

nature, was operative, and that in the case of nuclear matter,

$$\Gamma_{NM} \sim \frac{1}{2}\Gamma_{\Lambda}^{(0)} \qquad \text{Nuclear Matter, no correlations}$$

$$\Gamma_{NM} \sim \frac{1}{6}\Gamma_{\Lambda}^{(0)} \qquad \text{Nuclear Matter, with correlations.} \tag{5.140}$$

Thus, this "improved" calculation gave very different predictions from those of Dalitz. The ultimate arbiter is, of course, experiment, and the current state of affairs is shown in Figure 5.28. Here, the points with $A \sim 5$ represent old emulsion measurements, the $A = 16$ measurement is a 1976—low statistics, high background—LBL result [103], and the $A = 11$, 12 points are recently announced values from a BNL experiment by Grace et al. [104]. Clearly, the data indicate that the hypernuclear decay is slightly *faster* than that of the free Λ, in agreement with Dalitz's prediction. Yet the Adams calculation should be the better one. What's going on here?

In order to make sense out of this confused situation, our group (and others) undertook to redo the Adams evaluation with modern techniques.

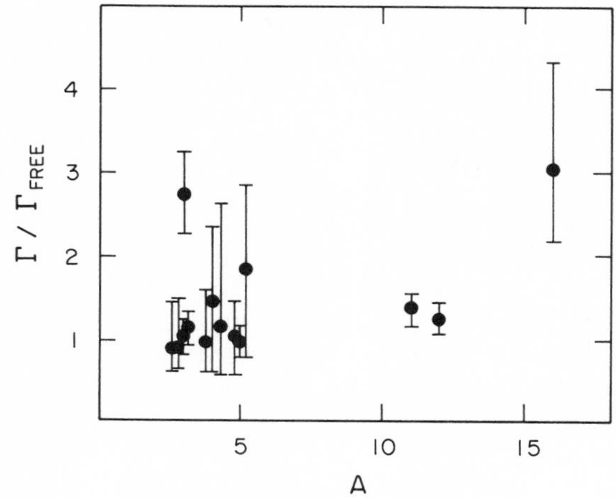

FIGURE 5.28. Plotted are experimental values for hypernuclear decay as a function of atomic mass number. Note that such rates tend to be of the order of but somewhat faster than that for decay of a free Λ. (J. F. Dubach, in *Weak and Electromagnetic Interactions in Nuclei*, ed. H. Klapdor, Springer, New York (1986)).

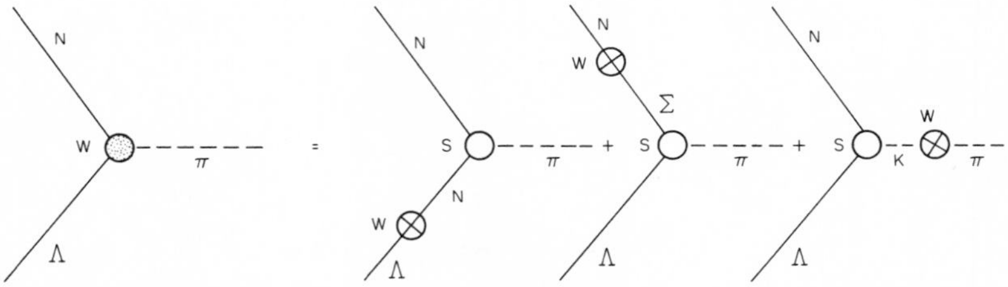

FIGURE 5.29. The pole model picture of the parity-conserving component of $\Lambda \to N\pi$ consists of three separate diagrams, as shown.

For example, the parity violating weak vertices

$$\langle NM|H_w|\Lambda\rangle, \qquad \langle NM|H_w|N\rangle$$

were evaluated using the same quark model—SU(6)$_w$—procedure that had been utilized, somewhat successfully, in the case of nuclear parity violation. The calculation is straightforward. However, there are many more vertices in this case because of the additional meson exchanges required for the hypernuclear analysis, as mentioned above. (A simplification compared to the nuclear parity violation case results from the feature that since the hypernuclear decay is a strangeness-changing process, only charged currents [W exchange] are involved.) Because I have already outlined these techniques in the previous section, I shall not review them here. Rest assured, however, that the calculations are quite tedious [105].

A new challenge arises in that the $\Lambda N \to NN$ transition can take place in either parity-violating *or* parity-conserving channels. Thus, parity-*conserving* $\Lambda \to NM$ and $N \to NM$ vertices are required. It is known that the baryon/meson pole model (Figure 5.29) gives a reasonable quantitative picture of the parity-conserving hyperon decay amplitudes: (although slightly different SU(3) parameters are required for the parity conserving vs. violating matrix elements) [106]. It seems prudent then to adopt the same model for the case of hypernuclear decay as indicated schematically in the diagrams in Figure 5.30. The two-particle weak parity *conserving* vertices are evaluated via current algebra—PCAC methods, which relate these amplitudes to experimental parity *violating* three-particle weak vertices [107]:

$$\begin{aligned} -\frac{i}{F_\pi}\langle N|[F^5,H_w]|\Lambda\rangle &\cong \langle N\pi|H_w|\Lambda\rangle \\ -\frac{i}{F_\pi}\langle \pi|[F^5,H_w]|K\rangle &\cong \langle \pi\pi|H_w|K\rangle. \end{aligned} \tag{5.141}$$

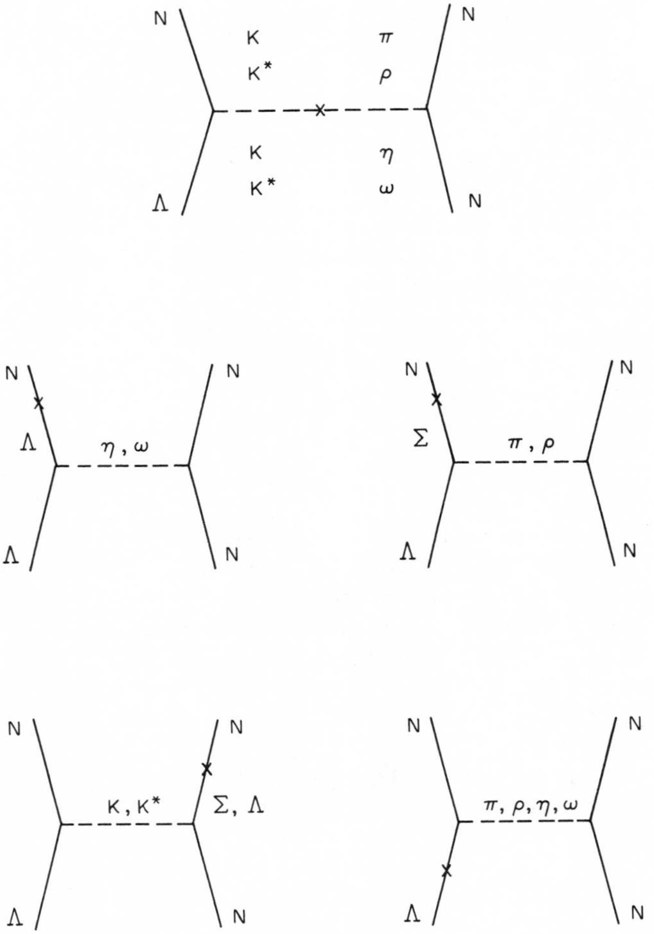

FIGURE 5.30. Many pole diagrams contribute to the parity-conserving component of weak hypernuclear decay.

Again, note that there are a large number of possible exchange diagrams that must be evaluated.

At this point, the Fourier transform is taken in order to convert these momentum space amplitudes to effective two-body coordinate space potentials. The resulting forms are rather complex. A typical piece is [108]

$$V^{(-)} = C_w C_s(U - 3Z)[T(\boldsymbol{\sigma}_\Lambda + \boldsymbol{\sigma}_N)T - (\boldsymbol{\sigma}_\Lambda - \boldsymbol{\sigma}_N)] \cdot \mathbf{f}_\pi^{(-)}(r), \quad (5.142)$$

where C_w, C_s are weak, strong coupling constants,

$$T = \frac{1}{4}(3 + \boldsymbol{\sigma}_N \cdot \boldsymbol{\sigma}_\Lambda), \qquad S = \frac{1}{4}(1 - \boldsymbol{\sigma}_N \cdot \boldsymbol{\sigma}_\Lambda)$$
$$U = \frac{1}{4}(3 + \boldsymbol{\tau}_N \cdot \boldsymbol{\tau}_\Lambda), \qquad Z = \frac{1}{4}(1 - \boldsymbol{\tau}_N \cdot \boldsymbol{\tau}_\Lambda) \tag{5.143}$$

are singlet, triplet operators in spin, isospin space, respectively [the Λ is considered as the isospinor $\binom{0}{1}$], and

$$\mathbf{f}_\pi^{(-)}(r) = \left[\frac{\mathbf{p}_1 - \mathbf{p}_2}{2m_N}, \frac{e^{-m_\pi r}}{4\pi r}\right] \tag{5.144}$$

has the typical Yukawa form.

Finally, we evaluate this effective potential between nuclear wave functions in order to make contact with the various observables. (The calculations sketched above are, though straightforward in principle, lengthy and time-consuming in practice. However, they are a perfect example of the merging of particle and nuclear physics described in the introduction. Only by the combined expertise of both particle *and* nuclear theorists is such work possible.) In the "nuclear matter" calculation, a Fermi gas model is employed, with equal numbers of neutrons and protons and a Fermi momentum of $p_F \cong 270$ MeV. The Λ is taken to be at rest and in a relative $^1S_0$ or $^3S_1$ state with respect to the nucleon with which it interacts. Of course, the final NN pair can be in a relative $^1S_0$, $^3S_1$, $^3D_1$, $^3P_0$, $^3P_1$, or $^1P_1$ state. The transition rate is calculated in the usual way:

$$\begin{aligned}\Gamma_{NN} &\sim \int_0^{p_F} d^3p \int d^3p_1 \int d^3p_2 \delta^4(p_f - p_i) \frac{1}{2} \sum_\sigma |\langle f|V|i\rangle|^2 \\ &= \sum_{\alpha,\beta} \Gamma(\beta \leftarrow \alpha).\end{aligned} \tag{5.145}$$

The results of the calculation are given in Table 5.5, which shows the ratio of hypernuclear decay rate for all the possible transition channels—$^1S_0 \to {}^1S_0$, $^3S_1 \to {}^1P_1$, etc.—divided by the free Λ decay rate, both with and without short-range correlations and with inclusion of various combinations of exchanged mesons. The results for pion-exchange only (colums 1 and 2) agree with Adams's assertion that correlations cut down the expected hypernuclear decay rate by a factor of about 3 or so. However, additional exchanges have a nonnegligible effect and the overall result is slightly *faster* than the free lamda-decay, rather than considerably

TABLE 5.5
CALCULATED VALUES FOR NONMESONIC HYPERNUCLEAR DECAY IN "NUCLEAR MATTER" $\Gamma_{NM}/\Gamma_{\text{FREE}}$

Transition	π No Cor	π Cor	$\pi + \rho$ Cor	$\pi,\rho,\eta,$ ω,K,K*	+"σ"
${}^1S_0 \leftarrow {}^1S_0$	.001	—	.001	.001	.004
${}^3P_0 \leftarrow {}^1S_0$	.156	.037	.052	.018	.018
${}^3P_1 \leftarrow {}^3S_1$	.312	.117	.113	.456	.456
${}^1P_1 \leftarrow {}^3S_1$	.468	.128	.100	.110	.110
${}^3S_1 \leftarrow {}^3S_1$	.010	.789	.589	.202	.200
${}^3D_1 \leftarrow {}^3S_1$	2.930	.751	.693	.444	.444
Total	3.890	1.830	1.550	1.230	1.230

Given are the calculated nonmesonic hypernuclear decay rates for "nuclear matter" broken down into channel-by-channel and exchange-by-exchange components.

slower as predicted by Adams. Also, it is interesting to note that the dominant decay in the pion-exchange-only scenario is into the $I = 0$ final state (${}^3S_1$ or ${}^3D_1$). Since an nn system must be purely $I = 1$, we predict that the π-exchange-only scenario requires

$$\frac{\Gamma_{NM}(\Lambda p)}{\Gamma_{NM}(\Lambda n)} \gg 1. \tag{5.146}$$

The experimental situation is shown in Figure 5.31, which reveals that the limited experimental data are consistent with

$$\frac{\Gamma_{NM}(\Lambda p)}{\Gamma_{NM}(\Lambda n)} \sim 1 \tag{5.147}$$

and quite inconsistent with the pion-exchange-only hypothesis The predictions of our calculation are given in Table 5.6, where we observe that inclusion of other meson exchanges brings the $\Lambda p/\Lambda n$ ratio down closer to experiment, *and* gives a value for the overall decay rate in good agreement with the new BNL results. Note also that another interesting test of the pion-exchange-only scenario exists if we examine the ratio of parity-violating to parity-conserving rates as shown in Table 5.6, column 1. Thus, a possible probe for the presence of additional meson exchanges would

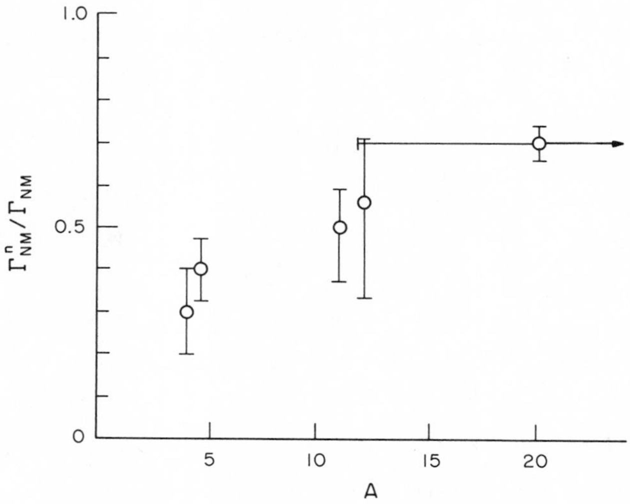

FIGURE 5.31. Plotted are experimental values of the ratio of neutron stimulated to total nonmesonic hypernuclear decay rates as a function of atomic mass number.

TABLE 5.6
CALCULATED VALUES FOR THE *PV/PC* RATIO AND FOR *p/n* STIMULATED HYPERNUCLEAR DECAY IN "NUCLEAR MATTER"

	Γ_{PV}/Γ_{PC}	$\Gamma_{\Lambda p\to np}/\Gamma_{\Lambda n\to nm}$
π No Cor	0.14	11.2
π Cor	0.18	16.6
$\pi+\rho$ Cor	0.21	13.1
π,ρ,η,ω,K,K*	0.90	2.9
+"σ"	0.90	2.9

Given are the values of the proton-to-neutron stimulation ratio and the comparison of parity-violating to parity-conserving rates for hypernuclear decay in nuclear matter.

be measurement of a parity-violating observable, such as the correlation between the direction of one of the outgoing nucleons and an initial Λ- or nuclear-polarization direction. This correlation should be maximal (small) under the all-meson (pion-only) scenario. Such an experiment has been proposed by a Japanese collaboration [109].

It is useful at this point to touch base with alternative calculations of hyper-nuclear decay. Thus,

1. McKellar and Gibson utilize a meson exchange approach with only π and ρ included and form factors at the vertices [110];
2. Cheung, Kisslinger, and Heddle employ a six-quark bag model wherein the weak interaction is divided into two sectors [111]:
   (a) π-exchange-only if $r > r_0 \cong 0.9\text{fm}$;
   (b) direct four-quark interaction if $r < r_0$.
3. Oset and Salcedo calculate the imaginary part of the hypernuclear self energy in a pion-exchange-only approximation, including the interactions of the pion with the nuclear medium [112].
4. Takeuchi, Takakai, and Bando employ a $\pi + \rho$ exchange picture, similar to that of McKellar and Gibson [113]. However, the model is applied to the calculation of the light hypernuclei ${}^5_\Lambda\text{He}$, ${}^4_\Lambda\text{He}$, and ${}^4_\Lambda\text{H}$.

Results of these calculations are shown in Table 5.7. Note that all agree that the Λ decay rate in nuclear matter is somewhat faster than that in free space, in contradistinction to Adams's prediction. (The reason for this

TABLE 5.7
COMPARISON OF VARIOUS THEORETICAL CALCULATIONS OF HYPERNUCLEAR DECAY $\Gamma_{NM}/\Gamma_{\text{FREE}}$

	$P \leftarrow S$	$D \leftarrow S$	Total
Dubach et al. (nuclear matter)	.584	.647	1.23
Adams	.380	.090	0.47
McKellar and Gibson	"small"		$2.30\ (1^{+2}_{-.4})$
Heddle and Kisslinger (nuclear matter)			3.00
Heddle and Kisslinger (${}^{12}_\Lambda\text{C}$)			1.28
Oset and Salcedo (nuclear matter)			2.00
Oset and Salcedo (${}^{12}_\Lambda\text{C}$)			1.50
Takeuchi, Takaki and Bando (${}^5_\Lambda\text{He}$)			0.45

Shown are the predictions made by various theoretical groups for hypernuclear decay rates.

TABLE 5.8
CALCULATED HYPERNUCLEAR DECAY RATES IN FINITE NUCLEI

		$\Gamma_{NM}/\Gamma_{\text{FREE}}$	Γ_{PV}/Γ_{PC}	$\Gamma_{\Lambda p}/\Gamma_{\Lambda n}$
${}^{5}_{\Lambda}\text{He}$	π (no corr.)	1.6	0.1	15.0
	π (with corr.)	0.9	0.1	19.0
	$\pi,\rho,K,\ldots$	0.5	0.8	2.1
${}^{12}_{\Lambda}\text{C}$	π (no corr.)	3.4	0.1	4.6
	π (with corr.)	3.0	0.1	5.0
	$\pi,\rho,K,\ldots$	1.2	1.1	1.6

Shown are the values of various hypernuclear decay quantities calculated for the case of finite nuclei.

difference is that Adams employed the wrong weak coupling constant as well as too strong a tensor correlation.) On the other hand, only Kisslinger et al. have quoted values for the $\Lambda p/\Lambda n$ or pv/pc ratios.

Of course, measurements performed within ${}^{12}\text{C}$ presumably should not be compared with calculations within nuclear matter. Thus we have undertaken shell model calculations in ${}^{12}_{\Lambda}\text{C}$ and ${}^{5}_{\Lambda}\text{He}$ hypernuclei. Results are given in Table 5.8 and are seen to be in reasonable agreement with the present limited experimental information.

We conclude that one can understand measurements of $\Gamma_{NM}/\Gamma_{\text{free}}$ although there remains considerable model dependence. However, we need more than lifetime values to distinguish between models. Particularly interesting would be measurements of $\Gamma(\Lambda p)/\Gamma(\Lambda n)$ and $\Gamma(pv)/\Gamma(pc)$, although other information such as nucleon spectra would also be welcome. The existence of a high-intensity kaon beam (as would be available in a kaon factory) would be helpful in this regard. It may also be possible to utilize a high-intensity CW electron facility such as CEBAF, which can produce hypernuclei via the $(e,e'K)$ reaction. In any case, it is clear that weak hypernuclear decay is a subject in its infancy and that our present theoretical optimism may well soon be tempered by experimental reality.

References

[1] See, e.g., L. D. Landau and E. M. Lifshitz, *Mechanics*, Pergamon, New York (1969), sec. 39.

[2] E. Merzbacher, *Quantum Mechanics*, Wiley, New York (1963), Ch. 15.

[3] G. Luders, Ann. Phys. (N.Y.) *2*, 1 (1957).
[4] C. R. Christenson et al., Phys. Rev. Lett. *13*, 138 (1964). A recent summary is provided by J. F. Donoghue, B. R. Holstein, and G. Valencia, Int. J. Mod. Phys. *A2*, 319 (1987).
[5] H. Burkhardt et al., Phys. Lett. *B206*, 169 (1988).
[6] T. D. Lee, Phys. Rev. *D8*, 1226 (1973), and Phys. Rept. *96*, 143 (1979); S. Weinberg, Phys. Rev. Lett. *37*, 657 (1976).
[7] S. Weinberg, Phys. Rev. Lett. *36*, 244 (1976); A. D. Linde, JETP Lett. *23*, 64 (1976).
[8] R. N. Mohapatra and J. C. Pati, Phys. Rev. *D11*, 566 (1975).
[9] M. Kobayashi and T. Maskawa, Prog. Theor. Phys. *49*, 652 (1973).
[10] L. Wolfenstein, Phys. Rev. Lett. *13*, 180 (1964).
[11] E. Blanke et al., Phys. Rev. Lett. *51*, 355 (1983).
[12] T. S. Bhatia et al., Phys. Rev. Lett. *48*, 227 (1982).
[13] K. Watson, Phys. Rev. *95*, 228 (1954); E. Fermi, Supp. Nuovo Cimento *2*, 58 (1955).
[14] C. E. Overseth and R. F. Roth, Phys. Rev. Lett. *19*, 391 (1967).
[15] S. W. Barnes, H. Winnick, K. Miyake, and K. Kinsey, Phys. Rev. *117*, 238 (1960).
[16] N. K. Cheung et al., Phys. Rev. *C16*, 2381 (1977); J. L. Gimlett et al., Phys. Rev. Lett. *42*, 354 (1979), and Phys. Rev. *C25*, 1567 (1980).
[17] B. R. Davis et al., Phys. Rev. *C22*, 1233 (1980).
[18] J. D. Jackson, S. B. Treiman, and H. W. Wyld, Phys. Rev. *106*, 517 (1957); C. E. Callan and S. B. Treiman, Phys. Rev. *162*, 1494 (1967); B. R. Holstein, Phys. Rev. *C5*, 1529 (1972).
[19] R. M. Baltrusaitis and F. P. Calaprice, Phys. Rev. Lett. *38*, 464 (1977); A. Hallin et al., Phys. Rev. Lett. *52*, 337 (1984); R. I. Steinberg et al., Phys. Rev. *D13*, 2469 (1976); B. Erozolimskii et al., Sov. J. Nucl. Phys. *28*, 48 (1978).
[20] J. D. Jackson, S. B. Treiman, and H. W. Wyld, ref. 18.
[21] M. B. Schneider et al., Phys. Rev. Lett. *51*, 1239 (1983).
[22] I. S. Altarev et al., Phys. Lett. *102B*, 13 (1981); J. M. Pendlebury et al., Phys. Lett. *136B*, 327 (1984); B. Heckel (priv. comm.).
[23] See, e.g., B.H.J. McKellar et al., Phys. Lett. *B197*, 556 (1987); I. B. Khriphlovich and A. Z. Zhitnitsky, Phys. Lett. *B109*, 490 (1982); E. Golowich and B. R. Holstein, Phys. Rev. *D26*, 182 (1982).
[24] P. Herczeg, Phys. Rev. *D28*, 200 (1983). X. He, B.H.J. McKellar and S. Pakvasa, Phys. Rev. Lett. *61*, 1267 (1988).
[25] G. Beall and N. Deshpande, Phys. Lett. *B132*, 427 (1983); I. I. Bigi and A. I. Deshpande, Phys. Rev. Lett. *58*, 1604 (1987).
[26] See, e.g., R. Jackiw in *Current Algebra and Anomalies*, by S. B. Treiman et al., Princeton Univ. Press, Princeton N.J. (1985).

[27] G. 't Hooft, Phys. Rev. Lett. *37*, 8 (1976); Phys. Rev. *D14*, 3432 (1976).
[28] R. J. Crewther et al., Phys. Lett. *88B*, 123 (1979), and *91B*, 487 (1980).
[29] See, e.g., E. M. Henley, Prog. Nucl. Part. Phys. *13*, 403 (1985).
[30] W. Haxton and E. M. Henley, Phys. Rev. Lett. *51*, 1937 (1983).
[31] V. V. Flambuam, I. B. Khriplovich, and C. P. Sushkov, Zh. Eksp. Teor. Fiz. *87*, 1521 (1984) [Sov. Phys. JETP *60*, 873 (1984)].
[32] J. F. Donoghue, B. R. Holstein, and M. J. Musolf, Phys. Lett. *B196*, 196 (1987).
[33] T. G. Vold et al., Phys. Rev. Lett. *52*, 2229 (1984). S. Lamoreaux et al., Phys. Rev. Lett. *59*, 2275 (1987); see also D. Schropp et al., Phys. Rev. Lett. *59*, 991 (1987).
[34] F. Raab in *Searches for New and Exotic Phenomena*, ed. T. Truong, Proc. Seventh Moriond Workshop (1987).
[35] L. I. Schiff, Phys. Rev. *132*, 2194 (1963).
[36] R. Peccei and H. Quinn, Phys. Rev. *D16*, 1791 (1977).
[37] F. Wilczek, in *How Far Are We From the Gauge Force*, ed. A. Zichichi, Plenum, New York (1985), p. 157.
[38] T. Donnelly et al., Phys. Rev. *D18*, 1607 (1978).
[39] F. P. Calaprice et al., Phys. Rev. *D20*, 2708 (1979).
[40] See, e.g., A. L. Hallin et al., Phys. Rev. Lett. *57*, 2105 (1986), and references therein.
[41] J. D. Bowman, in *Intersections Between Particle and Nuclear Physics*, ed. D. F. Geesaman, AIP Conf. Proc. No. 150, AIP, New York (1986), p. 1194.
[42] J. S. Altarev et al., Phys. Lett. *B102*, 13 (1981); V. P. Alfimenkov et al., Pisma Zh. Eksp. Teor. Fiz. *35*, 42 (1982) [JETP Letters *35*, 51 (1982)].
[43] J. D. Bowman, in *Weak and Electromagnetic Interactions in Nuclei*, ed. H. Klapdor, Springer, New York (1986), p. 633.
[44] R. C. Bohinski, *Modern Concepts in Biochemistry*, Allyn and Bacon, New York (1987), p. 383.
[45] R. P. Feynman, R. B. Leighton, and M. Sands, *The Feynman Lectures in Physics*, Addison-Wesley, Reading, Mass. (1964), sec. 52-4.
[46] R. P. Feynman et al., ref. 45, sec. 52-6.
[47] C. S. Wu et al., Phys. Rev. *105*, 1413 (1957).
[48] The seminal work on this subject is by F. Curtis Michel, Phys. Rev. *133*, B329 (1964).
[49] M. M. Nagels et al., Phys. Rev. *D12*, 744 (1975), and *D15*, 2547 (1977).

[50] G. Barton, Nuovo Cimento *19*, 512 (1961).
[51] D. E. Nagle et al. in *High Energy Physics with Polarized Beams and Polarized Targets*, Proc. of Third Int. Symp, Argonne, 1978, ed. by G. H. Thomas, AIP, Conf. Proc. No. 51, AIP, New York (1978), p. 224; J. M. Potter et al., Phys. Rev. Lett. *33*, 1307 (1974).
[52] R. Balzer et al., Phys. Rev. Lett. *44*, 699 (1980) and Phys. Rev. *C30*, 1404 (1984). S. Kistryn et al., Phys. Rev. Lett. *58*, 1616 (1987).
[53] P. von Rossen et al., in *Polarization Phenomena in Nuclear Physics–1980*, ed. by G. G. Ohlsen et al., AIP Conference Proc. No. 69, AIP, New York (1981), p. 1442.
[54] Quoted by D. E. Nagle et al., ref. 51.
[55] J. Lang et al., Phys. Rev. Lett. *54*, 170 (1985); R. Henneck et al., Phys. Rev. Lett. *48*, 725 (1982).
[56] V. Yuan et al., in *Intersections Between Particle and Nuclear Physics*, ed. D. F. Geesaman, AIP Conf. Proc. No. 150, AIP, New York (1980), p. 1184, and Phys. Rev. Lett. *57*, 1680 (1986). See also R. W. Harper et al., Phys. Rev. *D31*, 1151 (1985).
[57] N. Lockyer et al., Phys. Rev. Lett. *45*, 821 (1980).
[58] V. A. Knyazkov et al., Nucl. Phys. *A417*, 209 (1984).
[59] C. A. Barnes et al., Phys. Rev. Lett. *40*, 840 (1978).
[60] P. G. Bizzeti et al., Lett. Nuovo Cim. *29*, 167 (1980). M. Bini et al., Phys. Rev. Lett. *55*, 795 (1985).
[61] G. Ahrens et al., Nucl. Phys. *A390*, 486 (1982).
[62] H. C. Evans et al., Phys. Rev. Lett. *55*, 791 (1985).
[63] K. A. Snover et al., Phys. Rev. Lett. *41*, 145 (1978).
[64] E. D. Earle et al., Nucl. Phys. *A396*, 221 (1983).
[65] E. G. Adelberger in *Polarization Phenomena in Nuclear Physics*, AIP Conf. Proc. No. 69, AIP, New York (1981), p. 1367; E. G. Adelberger et al., Phys. Rev. *C27*, 2833 (1983).
[66] K. Elsener et al., Phys. Lett. *117B*, 167 (1982), and Phys. Rev. Lett. *52*, 1476 (1984).
[67] K. S. Krane et al., Phys. Rev. Lett. *26*, 1579 (1971), and Phys. Rev. *C4*, 1906 (1971).
[68] R. Wilson et al., in *The Investigation of Fundamental Interactions with Cold Neutrons*, ed. G. L. Greene, NBS Special Pub. No. 711, NBS, Washington (1986), p. 85.
[69] S. J. Shorka et al., St. Data *2*, 347 (1966); E. K. Warburton et al., Phys. Rev. *C20*, 619 (1979).
[70] E. G. Adelberger and W. C. Haxton, Ann. Rev. Nucl. Part. Sci. *35*, 501 (1985).
[71] K. Neubeck et al., Phys. Rev. *C10*, 320 (1974).

[72] B. A. Brown, W. A. Richter, and N. S. Godwin, Phys. Rev. Lett. *45*, 1681 (1980).
[73] B. Desplanques, *Proc. 8th Int. Workshop on Weak Interactions and Neutrinos*, Javea, Spain, World Scientific, Singapore (1983), p. 515.
[74] V. M. Lobashov et al., Sov. J. Nucl. Phys. *13*, 313 (1971); J. C. Vanderleeden and F. Boehm, Phys. Rev. *C2*, 748 (1970); E. Kuphal et al., Nucl. Phys. *A234*, 308 (1974); P. Bock and B. Jenachke, Nucl. Phys. *A160*, 550 (1971); E. D. Lipson et al., Phys. Rev. *C5*, 932 (1972).
[75] B. Desplanques, Nucl. Phys. *A316*, 244 (1979).
[76] V. P. Alfimenkov et al., ref. 42.
[77] M. Gari and J. H. Reid, Phys. Lett. *53B*, 237 (1974); J. K. Korner et al., Phys. Lett. *81B*, 365 (1979); V. M. Dubovik and S. V. Zenkin, Ann. Phys. (N.Y.) *172*, 100 (1986).
[78] B. Desplanques, J. F. Donoghue, and B.R. Holstein, Ann. Phys. (N.Y.) *124*, 449 (1980).
[79] F. C. Michel, ref. 31; H. Galic, B. Guberina, and D. Tadic, Z. Phys. *A276*, 65 (1976).
[80] J. F. Donoghue, Phys. Rev. *D15*, 184 (1977).
[81] H. J. Lipkin and S. Meshkov, Phys. Rev. Lett. *14*, 630 (1965); A. P. Balachandran, M. Gundzik, and S. Pakvasa, Phys. Rev. *D153*, 1553 (1967).
[82] B.H.J. McKellar and P. Pick, Phys. Rev. *D7*, 260 (1973).
[83] M. K. Gaillard and B. W. Lee, Phys. Rev. Lett. *33*, 108 (1974); G. Altarelli and L. Maiani, Phys. Lett. *B52*, 351 (1974); G. Altarelli, R. K. Ellis, L. Maiani, and R. Petronzio, Nucl. Phys. *B88*, 215 (1975).
[84] B. Desplanques, Nucl. Phys. *A335*, 147 (1980).
[85] W. C. Haxton, Phys. Rev. Lett. *46*, 698 (1981).
[86] C. Bennett, M. Lowry, and K. Krien, Bull. Am. Phys. Soc. *25*, 486 (1980).
[87] W. C. Haxton, ref. 85; E. G. Adelberger and W. C. Haxton, ref. 70.
[88] E. G. Adelberger et al., ref. 65.
[89] Note that this number is considerably smaller than originally calculated in ref. 85. Use of the beta-decay normalization yields

$$\frac{\langle O^-|V_{NN}^{(-)}|O^+\rangle^{\mathrm{expt}}}{\langle O^-|V_{NN}^{(-)}|O^+\rangle_{0,1\hbar\omega}} = 0.33 \pm 0.03,$$

in agreement with the effect found in mass 19, eq. (5.130).
[90] A careful discussion is presented in ref. 70.

[91] J. F. Donoghue and B. R. Holstein, Phys. Rev. Lett. *46*, 1603 (1981).

[92] Among the ongoing projects are a study of the $\Delta I = 1$ mixing of $J^p = 1^{\pm}$ levels at 11.2 MeV in ^{20}Ne by the Oxford group (L. K. Fifield et al., Nucl. Phys. *A394*, 1 [1983]), and improvement of the $A_\gamma(np)$ measurements at Grenoble (R. Wilson, ref. 68).

[93] E. G. Adelberger et al., *Annual Report of the Nuclear Physics Laboratory*, Univ. of Washington, Seattle (1987), p. 24.

[94] E. G. Adelberger et al., Phys. Rev. *C33*, 5 (1986).

[95] J. Birchall et al., TRIUMF Research Proposal E287—"Measurement of Parity Violation in $p - p$ Scattering." J. Birchall, Can. J. Phys. *66*, 530 (1988).

[96] M. Forte et al., Phys. Rev. Lett. *45*, 2088 (1980); B. Heckel et al., Phys. Lett. *119B*, 289 (1982), and Phys. Rev. *C29*, 2389 (1984).

[97] N. F. Ramsey, priv. comm.

[98] See, e.g., D. J. Millener in *Intersections Between Particle and Nuclear Physics*, ed. R. E. Mischke, AIP Conf. Proc. No. 123, AIP, New York (1984), p. 850.

[99] See, e.g., R. H. Dalitz, Nucl. Phys. *A354*, 101 (1981), and references therein.

[100] W. Cheston and H. Primakoff, Phys. Rev. *92*, 1537 (1953).

[101] M. M. Block and R. M. Dalitz, Phys. Rev. Lett. *11*, 96 (1963).

[102] J. B. Adams, Phys. Rev. *156*, 1611 (1967).

[103] K. J. Nield et al., Phys. Rev. *C13*, 1263 (1976).

[104] R. Grace et al., Phys. Rev. Lett. *55*, 1055 (1985).

[105] J. F. Dubach in *Weak and Electromagnetic Interactions in Nuclei*, ed. H. V. Klapdor, Springer-Verlag, New York (1986), p. 576.

[106] See, e.g., J. F. Donoghue, E. Golowich, and B. R. Holstein, Phys. Rept. *131*, 319 (1986), and references therein.

[107] J. F. Donoghue et al., Phys. Rev. *D21*, 186 (1980), and references therein.

[108] L. De la Torre, University of Massachusetts Ph.D. dissertation (1982).

[109] H. Ejiri et al., Phys. Rev. *C36*, 1435 (1987).

[110] B. Gibson and B.H.J. McKellar, Phys. Rev. *C30*, 322 (1984).

[111] C. Y. Cheung et al., Phys. Rev. *C27*, 335 (1983). D. P. Heddle and L. S. Kisslinger, Phys. Rev. *C33*, 608 (1986).

[112] E. Oset and L. L. Salcedo, Nucl. Phys. *A443*, 704 (1985).

[113] K. Takeuchi, H. Takaki, and H. Bando, Prog. Theor. Phys. Lett. *73*, 841 (1985). H. Bando, Prog. Theo. Phys. Suppl. *81*, 181 (1985).

Chapter 6

NEUTRINO PHYSICS

> Neutrinos, they are very small
> They have no charge and have no mass
> And do not interact at all.
>
> —JOHN UPDIKE

6.1 What's So Important about Neutrinos?

The neutrino, on the surface, seems to be a rather unimportant particle. It is neutral so it does not interact electromagnetically [1]. It is a lepton so it does not interact strongly. Thus it participates only in weak (and presumably gravitational) processes. Neutrino cross sections with matter are therefore tiny [$\sigma \sim G^2E^2 = 4 \times 10^{-38}(E/m_p)^2\ \text{cm}^2$], and in this sense the only critical role played by them appears to be in preserving conservation laws in various processes—e.g., spin, energy, momentum, etc.—which is why the existence of the neutrino was postulated in the first place. However, this relegation to second-class citizenship is quite inappropriate. In fact, the neutrino plays a role in physics far beyond that which is indicated by its tiny cross sections.

One aspect of this has to do with the ultimate fate of the universe. We know that at the present epoch the universe is expanding. But will it always continue to do so? Or will there come a time when gravitational attraction finally wins out and contraction sets in? In order to understand the role of the neutrino in answering this question, we assume that the universe is spatially homogeneous and isotropic. Now imagine a spherical piece of the universe of radius R and consider a small test particle of mass m which sits on the surface (cf. Figure 6.1). The mass contained within the spherical volume is given by

$$M = \frac{4}{3}\pi R^3 \rho, \tag{6.1}$$

where ρ is the mean energy density of the universe. Then the test mass m has potential energy

$$U = -m\frac{MG}{R} \tag{6.2}$$

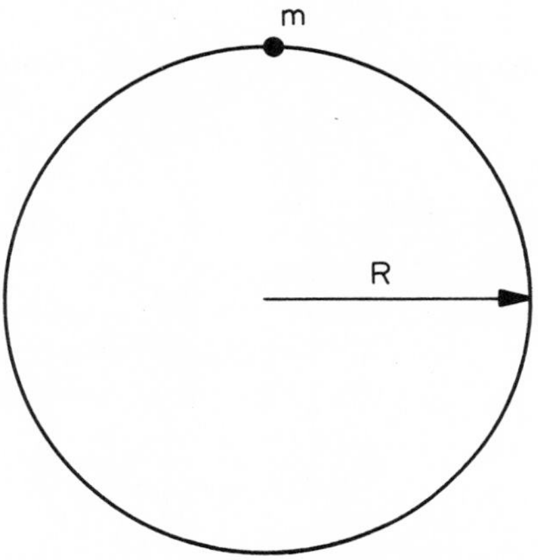

FIGURE 6.1. A test particle of mass m is situated on the surface of a spherical volume of radius R cut out of the universe.

and kinetic energy

$$KE = \frac{1}{2} mv^2 = \frac{1}{2} m \left(\frac{dR}{dt} \right)^2, \tag{6.3}$$

as seen by an observer at the center of the sphere. Now write the velocity as

$$v = HR, \tag{6.4}$$

where $H = \frac{1}{R}\frac{dR}{dt}$ is called the Hubble constant, and is found experimentally to be independent of R. The best current value is about [2]

$$H_{\text{exp}} \cong 100 \text{ km/sec/Megaparsec} \cong 3.0 \text{ cm/sec/lt yr} \tag{6.5}$$

(with considerable error bars). The total energy of the test particle then is given by

$$E_{\text{Tot}} = KE + PE = \frac{1}{2} mR^2 \left(H^2 - \frac{8}{3} \pi \rho G \right). \tag{6.6}$$

The condition for continued expansion is $E > 0$; if $E < 0$, the mass m is bound and can never completely escape to infinity. This defines a critical density [3]

$$\rho_c = \frac{3H^2}{8\pi G} \sim 2 \times 10^{-29}\ \mathrm{g/cm^3} \approx 4\,\frac{m_N}{\mathrm{m}^3}. \tag{6.7}$$

If $\rho > \rho_c$ there will come a time when the universe will begin to contract again, as shown in Figure 6.2. As to which situation describes our own universe, the evidence is uncertain. Analysis of the dynamics of gravitationally bound systems via the virial theorem together with measured luminosities yields the so-called visible mass density [4]

$$\frac{\rho_{\mathrm{VIS}}}{\rho_c} \sim 0.02. \tag{6.8}$$

However, analysis of the detailed rotational dynamics of spiral galaxies indicates that they are much more massive than their luminosity appears to indicate, with [5]

$$\frac{\rho_{\mathrm{EFF}}}{\rho_{\mathrm{VIS}}} \sim 20. \tag{6.9}$$

The origin or significance of this "dark matter" is unclear at present

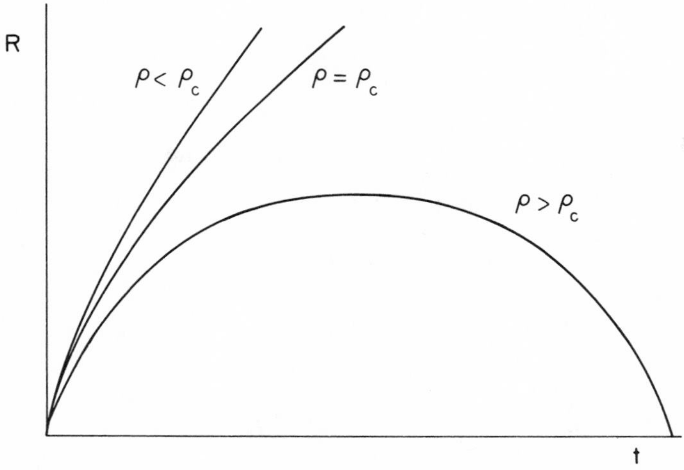

FIGURE 6.2. Plotted is the time evolution of the scale parameter R, which measures the scale of the universe, as a function of the cosmic mass density. Note that for $\rho \leq \rho_c$ the expansion continues forever, whereas if $\rho > \rho_c$ contraction will eventually take place.

(it could even be associated with neutrinos), and has been the source of a great deal of speculation [6]. Nevertheless, it does seem that the energy density of our universe is almost equal to the critical value needed for closure. Now, it is well known that a background black-body photon distribution at 2.7°K is left over from the early universe. The energy density of this photon gas is given by Stefan's law,

$$\rho_\gamma = 2\int \frac{d^3q}{(2\pi)^3}\frac{q}{e^{q/T}-1} = 4\sigma T^4, \tag{6.10}$$

where $\sigma = \pi^2/60$ is the Stefan-Boltzmann constant. In a similar fashion, a relic neutrino background remains from this era. The energy density associated with this distribution is given by

$$\rho_\nu = N_\nu \times \frac{7\pi^2}{120} T_\nu{}^4, \tag{6.11}$$

where N_ν is the number of "massless" ($m_\nu \ll T$) two-component neutrino species. (The constant in front of the neutrino expression differs from that which multiples the photon energy density because of the replacement of the Bose by the Fermi distribution in the sum over modes, viz.

$$\left.\int dz\,\frac{z^3}{e^z+1}\right/\int dz\,\frac{z^3}{e^z-1} = \left.\frac{7}{8}\right). \tag{6.12}$$

These neutrinos were originally in thermal equilibrium with electrons via reactions such as

$$e^+ + e^- \leftrightarrow \nu_e + \bar{\nu}_e, \qquad e^\pm + \nu_e \to e^\pm + \nu_e, \qquad e^\pm + \bar{\nu}_e \to e^\pm + \bar{\nu}_e. \tag{6.13}$$

However, once the reaction rate for such processes—$\sigma \cdot n \cdot v$ where n is the lepton number density—becomes smaller than the rate at which the temperature is decreasing—$-\dot{T}/T \sim \dot{R}/R = H$—the neutrinos drop out of thermal equilibrium with the charged leptons. This occurs at a temperature of about 10^{10}°K. From this point onward the neutrinos are essentially decoupled from the rest of the universe and cool down due to the expansion as

$$T \propto \frac{1}{a}, \tag{6.14}$$

where a is a scale parameter that measures the proper distance between fixed points on the fabric of the universe. Similar considerations hold for the photons that once were in thermal equilibrium with the electrons via the reaction

$$e^+ + e^- \leftrightarrow \gamma + \gamma.$$

Once the photons cool down to $T \gtrsim 10^{10}$°K, this reaction can proceed only to the right. Then e^+e^- annihilation into photons results in a heating process that raises the photon temperature with respect to that of the neutrinos by the factor [6]

$$\left(\frac{\rho_\gamma + \rho_{e^-} + \rho_{e^+}}{\rho_\gamma}\right)^{1/3} = \left(\frac{11}{4}\right)^{1/3}. \tag{6.15}$$

After combination of electrons and protons into hydrogen atoms at about 4000°K, the universe becomes transparent—these thermal photons are basically decoupled and cool down as the universe expands, reaching a temperature

$$T_\gamma = 2.7°\text{K} \tag{6.16}$$

in the present epoch. The corresponding relic neutrino temperature is then

$$T_\nu = 2.7°\text{K}\left(\frac{4}{11}\right)^{1/3} = 1.9°\text{K}. \tag{6.17}$$

Since cross sections are so small, these neutrinos are essentially unobservable. However, they have a nonnegligible energy density, which affects the expansion rate. Assuming masslessness, we find

$$\rho_\nu = \rho_\gamma \times \frac{7}{8} \times N_\nu \times \left(\frac{4}{11}\right)^{4/3} \approx 0.7\rho_\gamma \approx 3 \times 10^{-34}\ \text{g/cm}^3 \tag{6.18}$$

for $N_\nu = 3$ generations. However, if the neutrino had a mass—$\sum_i m_{\nu_i} \gtrsim 50$ eV—these cosmic neutrinos could themselves have sufficient energy density to close the universe! Thus the question of the possible existence of a nonzero neutrino mass is of considerable significance.

It is interesting also that neutrinos can tell us something about the number of quark generations. The argument again comes from astrophysics. First we note that the consistency of QCD requires that the sum of charges for all fundamental fermion fields must vanish [7].

$$\sum_{\text{Fermions}} Q_i = 0. \tag{6.19}$$

If this condition were to be violated, it can be shown that the existence of anomalies (cf. sec. 5.1) would destroy the renormalizability of the theory. Since we have

$$\begin{aligned} &\sum_i Q_i = -1 && \text{for each lepton generation} \\ &\sum_i Q_i = 3 \times \left(\frac{2}{3} - \frac{1}{3}\right) = +1 && \text{for each quark generation,} \end{aligned} \tag{6.20}$$

we see that the number of lepton generations must equal the number of quark generations:

$$N_l = N_q. \tag{6.21}$$

In order to determine N_l we note that in the very early universe—when $T > 10^{10}$°K—the reactions

$$\begin{aligned} n + e^+ &\leftrightarrow p + \bar{\nu}_e \\ p + e^- &\leftrightarrow n + \nu_e \\ n &\rightarrow p + e^- + \bar{\nu}_e \end{aligned}$$

took place so as to maintain thermal equilibrium with the ratio of neutron to proton densities given by [8]

$$n/p = \exp(-\Delta m/kT), \tag{6.22}$$

where $\Delta m = 1.3$ MeV is the neutron-proton mass difference. However, when $T \sim 1.5 \times 10^9$°K, the weak scatterings essentially cease in that the reaction rate drops below the expansion rate of the universe [9]. Thus the first two reactions above no longer take place at any appreciable rate, although the decay of neutrons to protons does continue. After the universe has cooled to 5×10^8°K, because of the electromagnetic processes

$$\begin{aligned} n + p &\rightarrow d + \gamma \\ d + d &\rightarrow \alpha + \gamma, \end{aligned}$$

essentially all the remaining neutrons end up within alpha particles, leaving extra protons left over. The ratio of the mass density of α-particles to the total mass density is

$$X = \frac{m_\alpha}{m_{\text{Tot}}} = \frac{4 \cdot \left(\frac{n}{2}\right)}{n + p} = \frac{2}{1 + \frac{p}{n}}, \tag{6.23}$$

where $\frac{p}{n}$ depends on the temperature at which the weak interactions freeze out, which in turn depends on the number of neutrino generations (since this, as we have seen, affects the overall energy density of the universe and therefore influences the expansion rate [10]). For the conventional three-generation model, we find

$$\frac{p}{n} \sim 7.3 \tag{6.24}$$

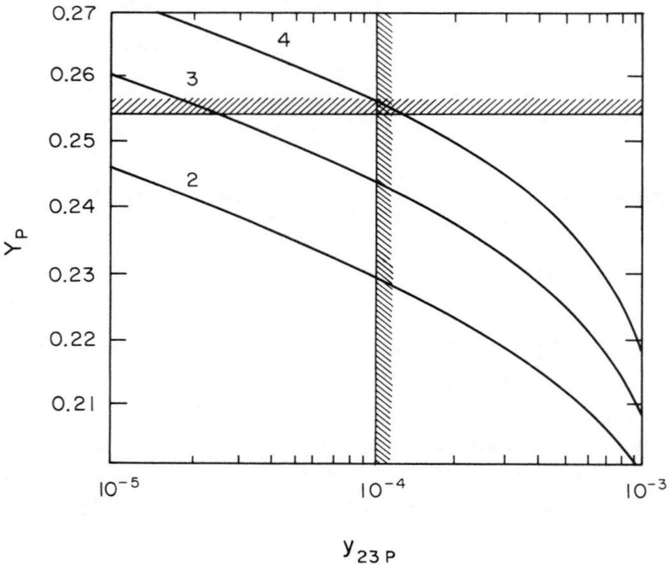

FIGURE 6.3. Plotted is the ^{4}He abundance Y_P vs. $D + {}^3$He abundance Y_{23P} as a function of the number of massless neutrino species. The allowed region—$Y_P < 0.254$, $Y_{23P} < 10^{-4}$—requires $N_\nu \leq 4$. © Gordon and Breach Science Publishers, S. A. [8]

and

$$X \simeq 0.24. \tag{6.25}$$

The situation for various numbers of neutrino generations is shown in Figure 6.3. If we believe conventional cosmology and the measurements of the primeval helium abundance, it is required that

$$N_l \leq 4, \tag{6.26}$$

which is an extremely significant result.

Although such cosmic speculations are interesting, let's get a bit more specific and ask what sort of evidence exists from earthbound experiments on the question of neutrino mass. We note in this regard that the validity of $SU(2)_L \times U(1)$ has nothing to say about this matter—failure to assign ν_R to any representation gives a massless neutrino, but this choice is not required by the group theory. The question of neutrino mass is an experimental one.

6.2 Neutrino Mixing

As we saw in Chapter 2, the standard model yields for the hadronic component of the charged weak current

$$J_\mu{}^{w}(\text{hadron}) = (\bar{u}\bar{c}\bar{t})\gamma_\mu(1+\gamma_5)U_{KM}\begin{pmatrix} d \\ s \\ b \end{pmatrix}, \tag{6.27}$$

where U_{KM} is the (unitary) KM matrix and relates the weak and mass quark eigenstates. Thus in terms of mass eigenstates the weak current is not generation-diagonal:

$$J_\mu{}^{w}(\text{hadron}) = \bar{u}\gamma_\mu(1+\gamma_5)(d\cos\theta_1 + s\sin\theta_1\cos\theta_3 + \ldots). \tag{6.28}$$

However, in the same model the leptonic component of the current was given as

$$\begin{aligned} J_\mu{}^{w}(\text{lepton}) &= (\bar{e}\bar{\mu}\bar{\tau})\gamma_\mu(1+\gamma_5)\begin{pmatrix} \nu_e \\ \nu_\mu \\ \nu_\tau \end{pmatrix} \\ &= \bar{e}\gamma_\mu(1+\gamma_5)\nu_e + \ldots. \end{aligned} \tag{6.29}$$

That is, there is no analog of U_{KM}—no unitary lepton mixing matrix. Why not? The answer is that the standard model assumes $m_{\nu_i} = 0$. Then even if the weak current had the mixed form

$$J_\mu{}^{\mathrm{w}}(\text{lepton}) = (\bar{e}\bar{\mu}\bar{\tau})\gamma_\mu(1 + \gamma_5)U_L \begin{pmatrix} \nu'_e \\ \nu'_\mu \\ \nu'_\tau \end{pmatrix} \tag{6.30}$$

with $U_L \neq 1$, we could simply *redefine* the identities of the neutrinos via

$$U_L \begin{pmatrix} \nu'_e \\ \nu'_\mu \\ \nu'_\tau \end{pmatrix} = \begin{pmatrix} \nu_e \\ \nu_\mu \\ \nu_\tau \end{pmatrix} \tag{6.31}$$

(which is allowed since all three neutrinos obey the same Dirac equation), yielding a weak current that *is* diagonal in lepton flavor.

On the other hand, if the neutrinos have mass, we have a term in the Lagrangian of the form

$$\mathscr{L}_m = m_{\nu_e}(\bar{\nu}'_{eR}\nu'_{eL} + \bar{\nu}'_{eL}\nu'_{eR}) + \dots . \tag{6.32}$$

Then, if we attempt to perform the left-handed rotation indicated above which diagonalizes the weak current, $\mathscr{L}_m$ would no longer be diagonal. Instead, we usually write the Lagrangians in terms of mass eigenstates, in which case the weak current is nondiagonal as in eq. (6.30). We would have the phenomenon of neutrino mixing, completely analogous to the quark mixing described by the *KM* angles. What would the experimental consequences of such mixing be?

To answer this question, imagine that there exist only *two* neutrino generations—ν_e and ν_τ—which are mixed by some angle θ. Then, in terms of *mass* eigenstates ν_1 and ν_2, we have

$$\begin{aligned} |\nu_e\rangle &= |\nu_1\rangle \cos\theta + |\nu_2\rangle \sin\theta \\ |\nu_\tau\rangle &= -|\nu_1\rangle \sin\theta + |\nu_2\rangle \cos\theta. \end{aligned} \tag{6.33}$$

If an electron-type neutrino (as defined by the weak interaction) is created at time $t = 0$, its wave function at some later time t will be given by

$$\begin{aligned} |\nu_e(t)\rangle &= |\nu_1\rangle \cos\theta e^{-iE_1t} + |\nu_2\rangle \sin\theta e^{-iE_2t} \\ &= |\nu_e\rangle(\cos^2\theta e^{-iE_1t} + \sin^2\theta e^{-iE_2t}) \\ &\quad + |\nu_\tau\rangle \sin\theta\cos\theta(e^{-iE_2t} - e^{-iE_1t}), \end{aligned} \tag{6.34}$$

where

$$E_1 = \sqrt{p^2 + m_{\nu_1}^2} \quad \text{and} \quad E_2 = \sqrt{p^2 + m_{\nu_2}^2} \tag{6.35}$$

are the neutrino energies. Thus the probability of "oscillating" into a ν_τ at time t is given by

$$P_\tau(t) = \sin^2 2\theta \sin^2 \frac{1}{2}(E_2 - E_1)t. \tag{6.36}$$

Now, since the neutrino mass is small (if not zero) compared to typical neutrino momenta, we can approximate

$$E_\nu = \sqrt{p^2 + m_\nu{}^2} \approx p + \frac{m_\nu{}^2}{2p} + \dots. \tag{6.37}$$

Then

$$P_\tau(t) \cong \sin^2 2\theta \sin^2 \frac{1}{4} \frac{\Delta m^2 L}{E_\nu}, \tag{6.38}$$

and this is the expression that is usually employed in analysis of neutrino oscillation experiments. (Note that instead of the time t we have written the expression in terms of the path length $L \approx t/c$.)

There are in general two types of mixing experiments. The first is termed a "disappearance" experiment. The idea is in this case to sit outside a reactor, from which a high flux of $\bar{\nu}_e$ is emitted. Although in general we do not know a precise value of the absolute flux, we *can* compare reaction rates induced by $\bar{\nu}_e$, e.g., $\bar{\nu}_e + p \to n + e^+$, at (massive) detectors located at differing distances L_1, L_2 from the reactor, as shown in Figure 6.4, top. If there exists *no* oscillation, the measured reaction rates will scale with neutrino flux $\sim \frac{1}{L^2}$. However, in the presence of neutrino mixing there would be superimposed on top of this $\frac{1}{L^2}$ fall-off an additional factor of $1 - P(t)$, and it is this modulation that signals the existence of the mixing phenomena.

The second type of oscillation experiment is called an "appearance" experiment. The point in this case is that at the end of, say, the LAMPF beam dump there are lots of neutrinos. However, $\bar{\nu}_e$'s are to a large extent missing. This is because the origin of the beam-dump neutrinos is primarily

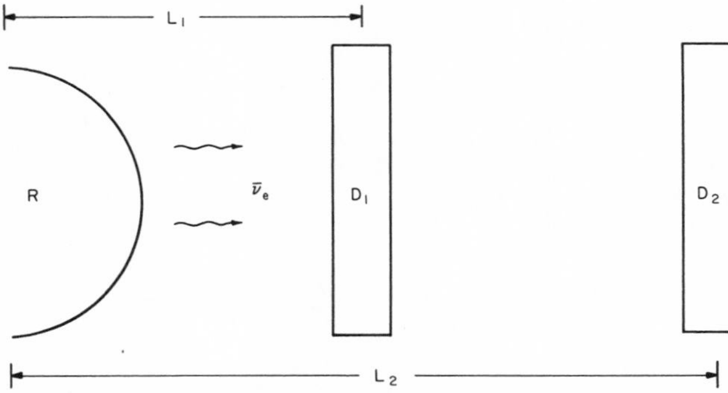

FIGURE 6.4a. A schematic drawing of a neutrino mixing experiment of the "disappearance" type, whereby a flux of reactor antineutrinos is monitored by detectors at different distances from the source. Deviation from a simple $1/r^2$ fall-off is sought.

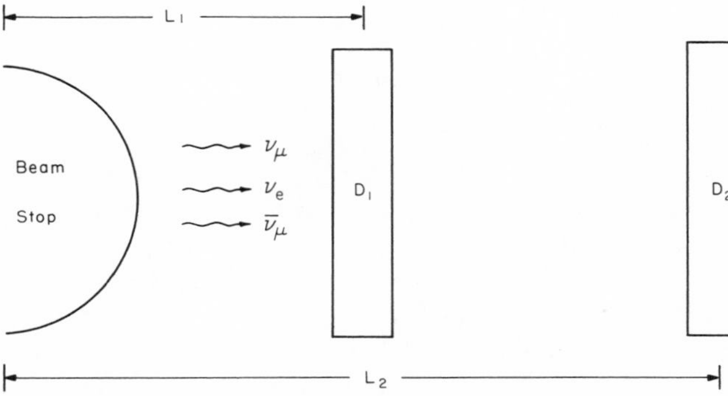

FIGURE 6.4b. A schematic drawing of a neutrino mixing experiment of the "appearance" type, whereby a set of detectors at various distances from a beamstop look for the presence of an anti-electron–neutrino flux. No signal is expected in the absence of mixing.

from pion beta decay,

$$\pi^+ \to \mu^+ \nu_\mu$$
$$\pi^- \to \mu^- \bar{\nu}_\mu.$$

The μ^+ is itself unstable and produces a flux of $\nu_e, \bar{\nu}_\mu$,

$$\mu^+ \to e^+ \nu_e \bar{\nu}_\mu.$$

However, the μ^- are predominantly captured before they can decay. For this reason, there are very few $\bar{\nu}_e$'s around after the beam stop. (Of course, there *are* some from the few μ^-,s which decayed before capture and from the rare (B.R. $\sim 10^{-4}$) decay $\pi^- \to e^-\bar{\nu}_e$. However, this is a small background.) We can then measure the rate for $\bar{\nu}_e$ interactions (as detected again via $\bar{\nu}_e + p \to n + e^+$) as a function of the distance from the beam stop (cf. Figure 6.4, bottom) as before. The rates will then depend upon $\frac{1}{L^2}$ times $P(t)$.

If no signal is observed in such experiments, it does not necessarily indicate that $P(t) = 0$. It could be that $\sin^2\theta$ is very small or that the path length L is simply not long enough for the oscillation to occur for a given value of Δm^2 and neutrino energy E_ν. It is conventional to present the results of such experiments on a two-dimensional diagram, with Δm^2 plotted against $\sin^2 2\theta$. Of course, regions of very small Δm^2 and θ can never be entirely ruled out. The level of sensitivity that can be achieved by such experiments can be seen from eq. (6.38),

$$\Delta m^2\Big|_{\min} \sim \frac{\langle E_\nu \rangle}{L_{\max}}. \tag{6.39}$$

Then, in a reactor experiment with $\langle E_\nu \rangle \sim 10$ MeV and a maximum distance of the detector from the source $L_{\max} \sim 100$ m, one can expect to be able to place a limit at the level

$$\Delta m^2 \sim \frac{10 \text{ MeV}}{100 \text{ m}} \sim 2 \times 10^{-2} \text{ eV}^2. \tag{6.40}$$

(It is interesting that reactor experiments, because of the low neutrino energies involved, have an advantage over their accelerator counterparts.) Of course, one can improve the sensitivity by increasing the distance of the detector from the source. However, a compromise must be reached between maximal distance and reasonable counting statistics.

Although occasional reports arise of a positive effect—most recently at the Bugey reactor [11]—the concensus, as summarized in Figure 6.5, is that at the present time no evidence exists for neutrino mixing. For reasonable values of the mixing angle—$\sin^2 2\theta > 0.1$—we find $\Delta m_\nu{}^2 < 0.1$ eV^2. An ongoing LAMPF $\bar{\nu}_\mu \to \bar{\nu}_e$ appearance experiment has also recently yielded limits in this ballpark [12].

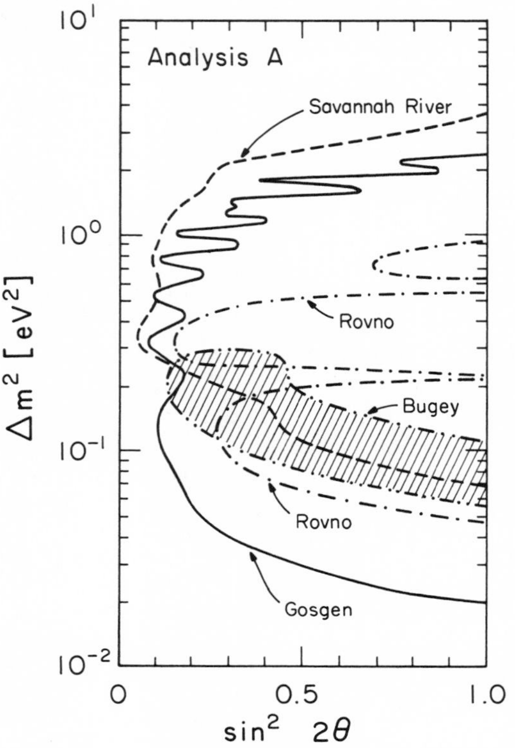

FIGURE 6.5. Shown is the region of Δm^2, $\sin^2 2\theta$ space excluded by various reactor neutrino mixing experiments. Note that while the Bugey collaborators claim a positive result, their allowed region is inconsistent with Gosgen and Savannah River data. (F. Boehm, Nucl. Phys. *A434*, 451 (1985)).

These direct neutrino mixing tests are experimentally extremely challenging because of the very small cross sections involved,

$$\frac{\sigma_w}{\sigma_{st}} \sim 10^{-15}, \tag{6.41}$$

and the corresponding low event rates. Massive detectors and large fluxes are required. Also, when completed, the evidence for neutrino mass is somewhat indirect: if mixing is found, then there does exist a neutrino mass, but what is actually observed is the mass *difference*. Finally, as mentioned above, if mixing is not detected, then this does not necessarily imply the nonexistence of neutrino mass. It could be that the mixing angle is simply very tiny. In any case, it is useful to seek more direct evidence, such as that obtained by analysis of semileptonic decay spectra.

6.3 Semileptonic Decay Spectra

If the neutrino has mass and/or if the weak and mass eigenstates of the neutrino are not identical, its effects will be strongly felt in the spectra of particles/nuclei decaying semileptonically. A simple example of this phenomena is observed in pion decay, $\pi^- \to \mu^- \bar{\nu}_\mu$. If the muon neutrino is a mixed state

$$|\bar{\nu}_\mu\rangle = \cos\theta|\bar{\nu}_1\rangle + \sin\theta|\bar{\nu}_2\rangle, \tag{6.42}$$

with, say, ν_1, ν_2 carrying different mass, then the μ^- spectrum would have the shape shown in Figure 6.6.

The detection of the second spike depends on the existence of a reasonable mixing angle as well as on precise detector resolution. Results from such experiments are shown in Figure 6.7. Being basically a high-energy physics experiment, the mass region ruled out is in the 10–100 MeV range.

Sensitivity to neutrino masses much less than 1 MeV is provided by careful analysis of beta-decay spectra, for which we can expect, in the massless neutrino case,

$$\frac{d\Gamma}{dE} \sim |M|^2(E_0 - E)^2 pEF(Z,E). \tag{6.43}$$

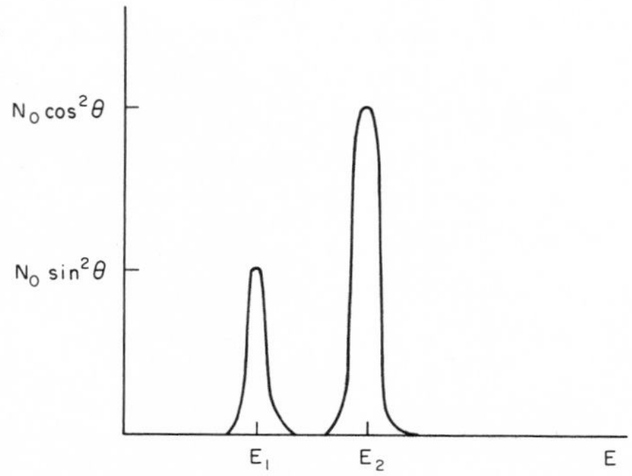

Figure 6.6. An idealized plot of the muon spectrum in the decay $\pi^- \to \mu^- \nu_\mu$, which would result in the case of two-component neutrino mixing.

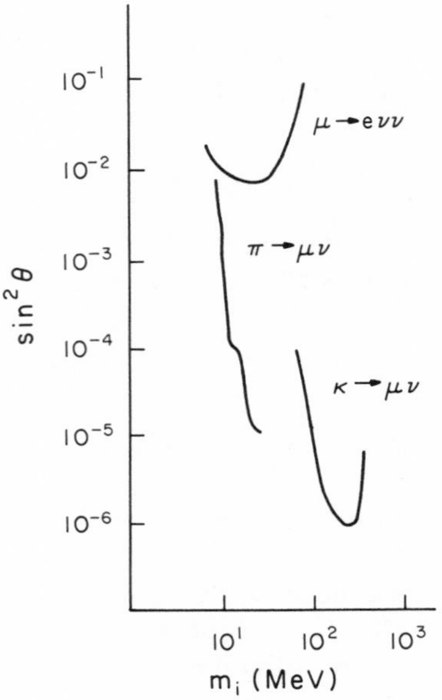

FIGURE 6.7. Shown are limits in m_i, $\sin^2\theta$ space obtained from spectral analysis of various elementary particle decays. Note that limits are in the 10 MeV–100 MeV range, as might be expected. (F. Boehm, Nucl. Phys. *A434*, 451 (1985)).

Then

$$R(E) \equiv \left(\frac{N(E)}{pEF(Z,E)}\right)^{1/2} \sim |M|(E_0 - E) \tag{6.44}$$

so that, to the extent that the nuclear matrix element is energy-independent, a plot of $R(E)$ vs. energy will be linear, as shown in Figure 6.8. This is, of course, just the familiar Kurie plot [13]. If neutrino mixing is present, on the other hand, the Kurie plot should have the form

$$R(E) = R_1(E)\cos^2\theta + R_2(E)\sin^2\theta, \tag{6.45}$$

whereas, neglecting mixing, if the neutrino has a mass

$$\frac{d\Gamma}{dE} \sim |M|^2 pE(E_0 - E)[(E_0 - E)^2 - m_\nu{}^2]^{1/2}, \tag{6.46}$$

which is distinguished from the $m_\nu = 0$ case by measurements near the endpoint—$E_0 \sim E$—cf. Figure 6.9.

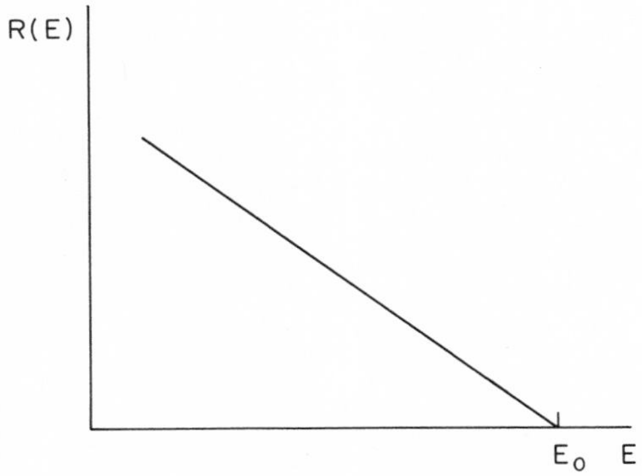

FIGURE 6.8. An idealized Kurie plot for the case of vanishing neutrino mass.

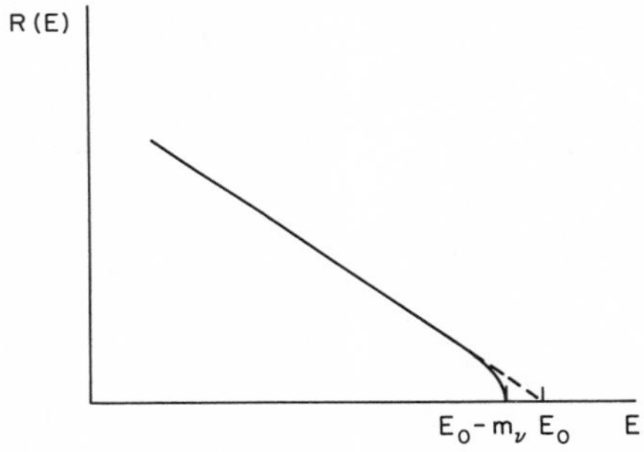

FIGURE 6.9. An idealized Kurie plot for the case of nonzero neutrino mass.

These are difficult measurements, requiring enormous experimental skill, and interestingly both types of experiments have recently reported a signal. In the former case, Simpson at Guelph has measured the decay spectrum from the decay of tritium, ${}^3H \to {}^3He + e^- + \bar{\nu}_e$, and has reported evidence for neutrino mixing, with [14]

$$\begin{aligned} \sin^2\theta &\sim 0.03 \\ m_{\nu_1} &\sim 0 \\ m_{\nu_2} &\sim 17.1 \text{ KeV}. \end{aligned} \tag{6.47}$$

As seen in Figure 6.10, the Kurie plot for electron energies $T_\beta < 1.5$ KeV reveals a systematic deviation from the simple ($\theta = 0$, $m_\nu = 0$) straight-line value—note here $T_0 = 18.6$ KeV.

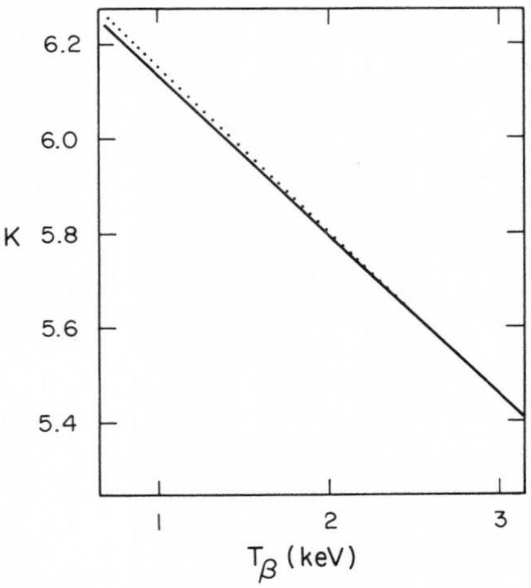

FIGURE 6.10. Shown is the Kurie plot obtained by Simpson [14] for tritium beta decay. Note the deviation from linearity that occurs at $T_\beta \approx 1.5$ KeV.

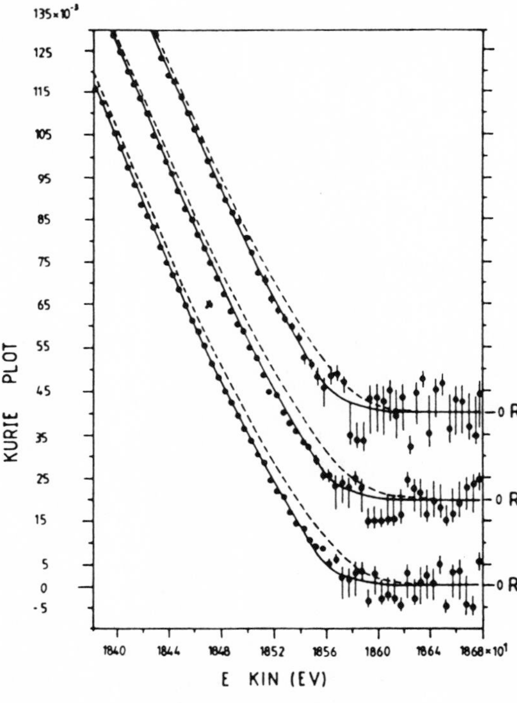

FIGURE 6.11. Shown is experimental data from the Soviet tritium beta-decay experiment [15]. The dotted line represents a zero neutrino mass fit, while the solid line represents the expected spectrum if $m_\nu \sim 30$ eV. The three lines represent three different sources.

In addition, in an experiment involving tritium bound in a valine ($C_5H_{11}O_2N$) molecule, careful analysis of the Kurie plot near the endpoint by an ITEP group has revealed a systematic deviation [15], as shown in Figure 6.11. Here the dotted line indicates a fit with $m_\nu = 0$, while the solid line represents the expected shape with an assumed mass of 30 eV. From careful statistical analysis of such fits, the Soviet group claims to find a nonzero value:

$$17 \text{ eV} < m_\nu < 40 \text{ eV}. \tag{6.48}$$

However, both results are controversial and have stimulated a great deal of related work. A number of groups have looked at the spectra of the Gamow-Teller transition, ${}^{35}S \rightarrow {}^{35}Cl + e^- + \bar{\nu}_e$, and have reported *no* evidence for a 17 KeV neutrino branch [16]. The result of a Princeton experiment is shown in Figure 6.12. The dashed curve shows the predicted ratio for Simpson's 17 KeV neutrino with a 3% branching ratio. The solid line represents a simple one-component massless neutrino. Obviously, there is no evidence for the existence of the massive 17 KeV neutrino, and one finds a limit on a possible mixing parameter,

$$\tan^2\theta < 0.3\%. \tag{6.49}$$

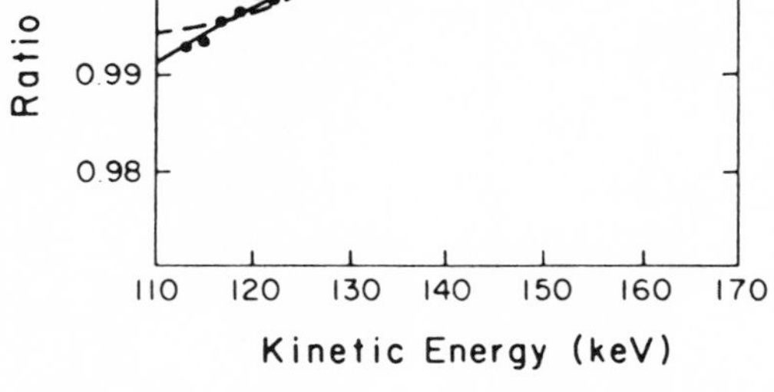

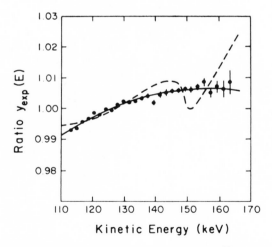

FIGURE 6.12. Shown is the spectral shape in the beta decay of ${}^{35}S$ as measured by the Princeton group [16]. The dotted line indicates what would be expected assuming the existence of Simpson's hypothesized 17 KeV neutrino with a 3% branching ratio.

The other experimental groups are in agreement, and the present view is that the origin of Simpson's effect is not understood, although it is possible that the anomaly may be partially due to atomic physics effects [17].

The validity of the ITEP experiment has also been called into question by Los Alamos and Zurich measurements of the tritium spectrum. The Kurie plot found by the LANL group is shown in Figure 6.13 and indicates no evidence for a deviation from linearity. Analysis of this data yields the upper bound [18]

$$m_\nu < 27 \text{ eV @ } 95\% \text{ C.L.}, \tag{6.50}$$

while the Zurich group reports from a similar experiment [19]

$$m_\nu < 18 \text{ eV}. \tag{6.51}$$

Although not strictly in disagreement with the reported Soviet number, there is also no indication of a nonzero neutrino mass.

The Russian experiment has been criticized in the literature on two fronts. First, their measured spectrum is critically dependent upon their resolution, which tends to decrease the slope of the Kurie plot near the endpoint. Thus the measured spectrum does not possess the striking downturn near the endpoint expected from Figure 6.9 and in fact turns the opposite direction, yielding a neutrino mass value that is critically dependent upon the resolution in this region [20]. The resolution has been measured using conversion electrons, but important questions remain [21].

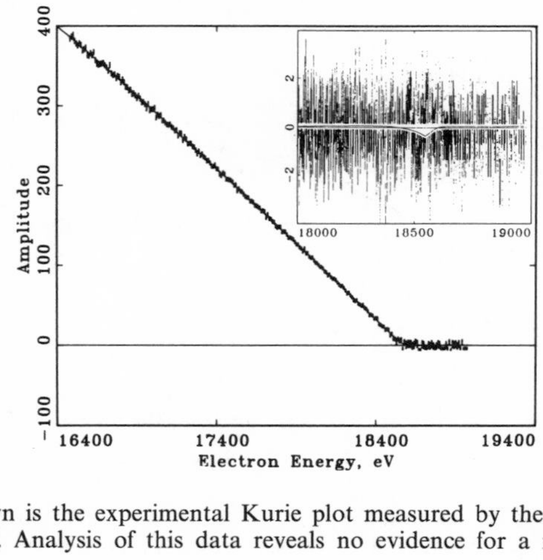

FIGURE 6.13. Shown is the experimental Kurie plot measured by the Los Alamos collaboration [23]. Analysis of this data reveals no evidence for a nonvanishing neutrino mass.

A second problem concerns the fact that the tritium source was part of a valine molecule. The measured spectrum involves a weighted average over transitions to not only the ground state but also to excited states of the atomic, molecular and even of the crystalline system, all of which must be calculated. In this regard the use of molecular tritium by the Los Alamos group is to be preferred [22].

In any case, other experiments involving ^3H are underway at a number of laboratories around the world, and in a year or two several very precise numbers should be forthcoming on the possible existence of a non-zero mass of the electron neutrino with a sensitivity of 10 eV or so [23]. Another promising approach involves analysis of the spectrum of radiative electron capture $e^- + {}^AN \rightarrow \nu_e + {}^AN' + \gamma$. However, at present the best limit is only at the 500 eV level [24].

In the midst of all this experimental activity a serendipitous occurrence has yielded the opportunity to make neutrino mass "measurements" of almost equal precision, although with some attendant model dependence. This occurrence was, of course, the supernova explosion that was detected on February 23, 1987 (cf. Figure 6.16). That this was serendipitous is clear for a number of reasons. First, and most important, it is the first supernova observed in our vicinity in nearly 400 years, the last such occurrence having been noted by Kepler in 1604! Second, the supernova was observed optically by a Canadian astronomer who happened to be looking that night (in Chile) at that particular region of the sky, the Large Magellenic Cloud. Third, two large tanks of water (one at Kamiokande, in Japan, and the other at IMB, in Ohio) instrumented for the detection of possible decay of the proton were in operation at the time and observed the event via a burst of anti-neutrinos, emitted by the supernova, which passed through these tanks, initiating the reaction $\bar{\nu}_e + p \rightarrow e^+ + n$, which was seen via the resulting Cerenkov radiation associated with the e^+ motion.

Before discussing how such mass measurements are achieved, I will now describe briefly the (type II) supernova process itself. A typical star, such as our sun, generates energy by nuclear fusion, whereby hydrogen is converted to helium. The process itself involves a long chain of reactions, as described in greater detail later in this chapter. However, the net effect is described by the simple sequence, $2e^- + 4p \rightarrow {}^4\text{He} + 2\nu_e + 26.7$ MeV. Such a star, with a core temperature $\sim 10^7$°K is said to reside on the "main sequence" of a Hertzsprung-Russell diagram plotting luminosity versus temperature (Figure 6.14). Depending on its mass, the star can spend various lengths of time in this region, with several billion years being not atypical. Eventually, however, the hydrogen contained in the core begins to dwindle. When this happens, gravitational attraction contracts the core, increasing the temperature and therefore the reaction rate at which the

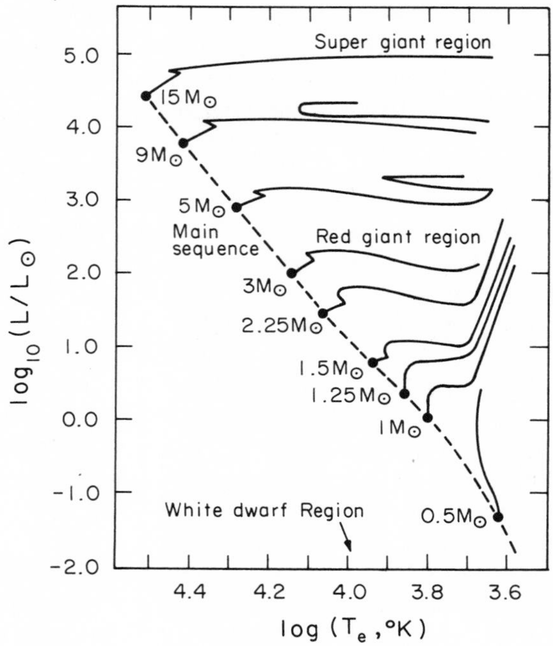

FIGURE 6.14. Shown is a Hertzsprung-Russell diagram plotting stellar luminosity vs. surface temperature. During the hydrogen-burning stage, which typically lasts several billion years, a star resides in the main sequence region, moving off this line as the hydrogen fuel becomes exhausted. (E. Commins, *Weak Interactions*, © 1973 McGraw-Hill)

remaining hydrogen is "burned." The increased rate of energy emission causes the outer layers of the star to expand, resulting in a decreased surface temperature. The star then becomes a so-called red giant, having a much larger size and luminosity than before, but a cooler surface temperature. The core, of course, continues to heat up, as the proton supply runs low and contraction continues. When the temperature reaches about 10^{8}°K helium burning, $3\,{}^4\text{He} \rightarrow {}^{12}\text{C} + \gamma$, commences ($2\,{}^4\text{He} \rightarrow {}^8\text{Be} + \gamma$ cannot occur since ^{8}Be is unstable), temporarily halting the gravitational contraction. The higher temperature is required in order to overcome the Coulomb repulsion between ^{4}He nuclei, which is four times stronger than that between two protons. This helium-burning phase of stellar evolution is much shorter than its hydrogen burning counterpart—indeed, the core is much hotter and the reaction rate much higher. What eventually happens to such a helium burning star depends on its mass. If the star is relatively light—less than about 3 or 4 solar masses—then as the ^{4}He supply runs low and the core shrinks under the ever-present influence of gravity, eventually the Pauli exclusion principle associated with the electron "gas" within the core wins out. Collapse is halted and the gravitational attraction is stabilized by the outward pressure of this degenerate electron gas. Through a mechanism still not completely understood, the envelope of gaseous material surrounding the core is shed (producing a planetary nebula), leaving a white dwarf star, containing basically carbon nuclei and the electron gas. For stability the mass of this white dwarf must be less than 1.4 solar masses—the so-called Chandrasekhar limit [25]. After this, the star cools by radiating its energy into space and eventually becomes an invisible black dwarf.

What occurs if the star is much heavier than 3 or 4 solar masses is more interesting. In this case the degeneracy pressure of electrons due to the exclusion principle is insufficient to halt further gravitational contraction, and as the core compresses and heats up further, additional thermonuclear reactions are initiated. Thus at about 6×10^{8}°K carbon itself begins to burn via

$$\begin{aligned} {}^{12}\mathrm{C} + {}^{12}\mathrm{C} &\to {}^{20}\mathrm{Ne} + {}^{4}\mathrm{He} \\ &\to {}^{24}\mathrm{Mg} + \gamma \\ &\to {}^{16}\mathrm{O} + 2\,{}^{4}\mathrm{He} \\ &\text{etc.} \end{aligned}$$

The length of time spent in this mode—say, 1000 years—is even shorter than that spent in the helium phase, and as the carbon supply begins to run low, gravitational contraction heats the core even further, allowing neon burning

$$\begin{aligned} {}^{20}\mathrm{Ne} + \gamma &\to {}^{16}\mathrm{O} + {}^{4}\mathrm{He} \\ {}^{20}\mathrm{Ne} + {}^{4}\mathrm{He} &\to {}^{24}\mathrm{Mg} + \gamma \\ &\text{etc} \end{aligned}$$

to commence at about 1×10^{9}°K. As the evolution continues and the temperature reaches about 1.5×10^{9}°K, oxygen burning is initiated, producing a wide range of final products:

$$\begin{aligned} & {}^{32}\mathrm{S} + \gamma \\ & {}^{28}\mathrm{Si} + {}^{4}\mathrm{He} + \gamma \\ {}^{16}\mathrm{O} + {}^{16}\mathrm{O} \to\ & {}^{31}\mathrm{S} + n + \gamma \\ & {}^{24}\mathrm{Mg} + 2\,{}^{4}\mathrm{He} \\ & \text{etc.} \end{aligned}$$

Finally, at a temperature of about 3×10^{9}°K, silicon burning is started, again producing a varity of heavier elements, among which the most prevalent (and important) is Fe. As is well known, nuclei in this region of the periodic table have the maximum binding energy per nucleon; thus, as these heavy elements are piled up in the central core, no further thermonuclear reaction is possible. The configuration of the star at this point consists of a very hot and dense core containing primarily Fe, surrounded by a mantle consisting mainly of Si at somewhat lower temperature and pressure. Further out from the center exist additional shells with Mg, Ne, etc., until we reach the surface, where an envelope of hydrogen and helium still exists. Thus we have the so-called onion skin picture shown in Figure 6.15.

One obvious question here is that if no further thermonuclear reactions are taking place at the core, what keeps it from collapsing? The answer is that the stability is due to electron degeneracy pressure, just as in the case of the white dwarf. However, now there is an important difference. As Si burning continues, more and more Fe ash is deposited in the core. Eventually the Chandrasekhar limit is exceeded and catastrophe ensues. Unable to sustain the tremendous inward crush that is due to gravity, the

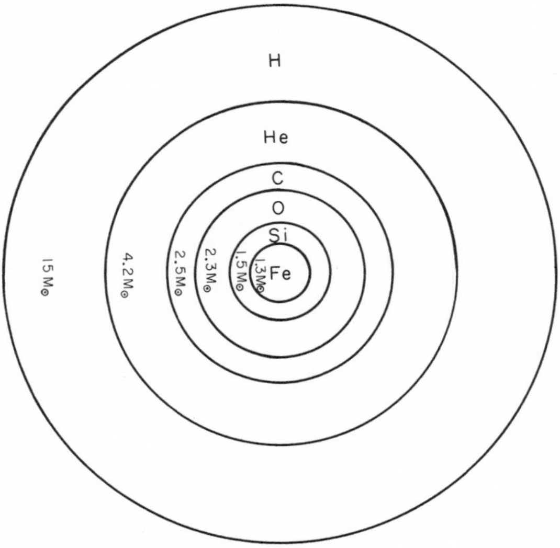

FIGURE 6.15. Shown is a schematic picture of a star in the supernova stage. Note the increasing atomic mass number as one proceeds toward the stellar center. The hydrogen that once filled the interior is now confined to the surface.

core region collapses and begins rapidly to contract, resulting in a sharp rise in the temperature. Photonuclear reactions dissociate the remaining nuclei into their fundamental proton and neutron constituents, and the protons and electrons themselves combine to form even more neutrons, emitting an electron-type neutrino in the process: $p + \bar{e} \rightarrow n + \nu_e$. This "neutronization" results in a core region containing primarily neutrons, and eventually one reaches the stage where the central region consists of about one solar mass of neutrons contained within a radius of about 10 km. At this point the core is at nuclear density and the degeneracy pressure due to the Pauli exclusion effect for these neutrons abruptly halts the collapse, as the core becomes virtually incompressible. Of course, stellar matter from outside this core region is no longer supported and is still rushing toward the center. When this material comes in contact with the now rigid neutron core, a huge rise in temperature and pressure develops, causing the inflowing material to "bounce" and begin to rush away toward the surface. In this fashion a shock wave develops, shedding much of the outer gaseous stellar medium and leaving a neutron star at the center.

The dominant mechanism for carrying energy away from the supernova is the emission of neutrinos. The first burst is, as mentioned above, associated with the neutronization process and occurs during the very brief (a fraction of a second) period of core collapse. A second flux of neutrinos is associated with the fact that the collapsing core involves very high temperatures ($T \sim 6 \times 10^{10}$°K $\sim$ 5 MeV) so that neutrinos are thermalized and produced by many means:

(1) Pair annihilation

$$e^+ + e^- \rightarrow \nu_e + \bar{\nu}_e$$

(2) Photonuclear production

$$\gamma + A \rightarrow A + \nu_e + \bar{\nu}_e$$

(3) The Urca reaction

$$e^- + (Z,N) \rightarrow \nu_e + (Z - 1, N + 1)$$

etc

The core densities are so high that the diffusion process whereby the neutrinos make their way out of the core region takes several seconds or so. (This large energy flux may well play an important role in the shedding of the envelope.) The precise process by which such a supernova explosion occurs is obviously a very complex one, and detailed numerical studies are required. Recent calculations seem to be able to simulate a supernova explosion, but the theory of this subject is still in its infancy, and much more work is needed before we will have achieved a real understanding [26].

The supernova 1987a, which was seen on 23 February of that year, occurred in the Large Magellenic Cloud about 170,000 years ago. Besides the optical signal that is perhaps the most spectacular evidence for this event (Figure 6.16) there was also, according to arguments presented above, a pulse of $\nu,\bar{\nu}$ generated at the same time. They have been traveling through space (at the speed of light if neutrinos are massless) for 170,000 years and struck neutrino detectors on opposite sides of the globe no more than about three hours before the arrival of the optical signal. The neutrino spectrum itself is expected to be essentially thermal, with a characteristic temperature of the order 5 MeV or so, and spread out over a period of time associated with the diffusion process. This is roughly what is observed

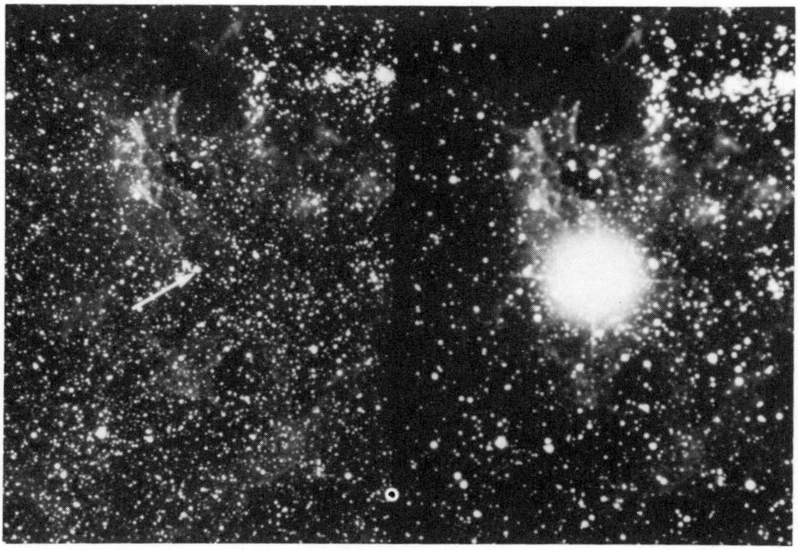

FIGURE 6.16. Shown (before and after) is the supernova 1987a, which appeared on the night of 27 February 1987 in the Large Magellenic Cloud. Although not in the Milky Way, this supernova is only 170,000 light years distant and is the first to be observed in our vicinity in 350 years. © 1987 Anglo-Australian Telescope Board.

by the two experimental groups—a burst of events with energies of the order tens of MeV over a time period of several seconds [27].

So how does sensitivity to neutrino mass arise? Simply speaking, if a neutrino has mass, its velocity will depend on that mass via

$$v = \frac{p_\nu}{\sqrt{p_\nu{}^2 + m_\nu{}^2}} \simeq 1 - \frac{m_\nu{}^2}{2p_\nu{}^2} \simeq 1 - \frac{m_\nu{}^2}{2E_\nu{}^2}. \tag{6.52a}$$

Then the time of arrival of energetic neutrinos will be earlier than their lower energy counterparts. The difference in arrival times is roughly

$$\frac{\delta t}{t} \sim \frac{\delta v}{v} \sim \frac{m_\nu{}^2}{E_\nu{}^2}\frac{\delta E_\nu}{E_\nu}. \tag{6.52b}$$

The eleven events seen by the Kamiokande detector and eight seen by the IMB group have an energy spread $\delta E_\nu \sim 10$ MeV and arrive over roughly a ten-second span of time. Thus we find in this very simple scenario,

$$m_\nu \lesssim E_\nu\left(\frac{\delta t}{t}\frac{E_\nu}{\delta E_\nu}\right)^{1/2} \sim 10\ \text{MeV}\left(\frac{10\ \text{sec}}{10^{13}\ \text{sec}}\right)^{1/2} = 10\ \text{eV}. \tag{6.53a}$$

Of course, this analysis is simplistic. A careful model for the time and energy distribution is required in order to yield a realistic limit. When this is done we find [28]

$$m_\nu \lesssim 20\ \text{eV}. \tag{6.53b}$$

Actually, we can get much more [28]. From the observed neutrino flux we can limit the number of neutrino generations to $\lesssim 6$, although astrophysical bounds described above are more precise. Because optical and neutrino signals arrived within several hours of one another despite a time delay of several months induced by general relativistic considerations, we have verification that photons and neutrinos travel identical spacetime trajectories, and therefore that neutrinos indeed interact gravitationally. Other bounds are provided on possible axion or majoron emission, etc. It is remarkable that a handful of neutrinos observed in a pulse of several seconds from a supernova allows a limit to be placed on the neutrino and other particles that are competitive with those gained from years of high-statistics, high-precision weak interaction work!

6.4 Double Beta Decay

A third approach to the problem of neutrino mass, albeit an indirect one, is based on a search for the neutrinoless double beta decay reaction. The key to understanding the double beta decay process lies in the existence of the pairing force, because of which even-even nuclei, whose spins are all paired to zero, are bound more strongly than odd A neighbors. In such a case it is possible for a simple beta decay transition to be kinematically forbidden, while decay via emission of two beta particles is not. An example is ^{82}Se as shown in Figure 6.17, wherein we observe that the ^{82}Se nuclide is stable with respect to beta decay to ^{82}Br, but that double beta emission to ^{82}Kr is permitted. Many similar examples of this phenomenon exist in nature, such as ^{238}U, ^{232}Th, ^{128}Te, ^{130}Te, etc., and many of these have been probed for the existence of the double beta phenomenon, as described below.

The important question is the reaction mechanism by which the double beta transition takes place. This depends on whether the neutrino is a Majorana or Dirac particle, i.e., whether the charge conjugate state of the neutrino is identical to (Majorana) or different from (Dirac) the neutrino itself:

$$\begin{aligned} C|\nu\rangle &\equiv |\bar{\nu}\rangle = |\nu\rangle \qquad \text{Majorana} \\ C|\nu\rangle &\equiv |\bar{\nu}\rangle \neq |\nu\rangle \qquad \text{Dirac} \end{aligned} \tag{6.54}$$

Obviously then, a Majorana particle must be neutral and cannot carry any internal quantum numbers, but more about that later.

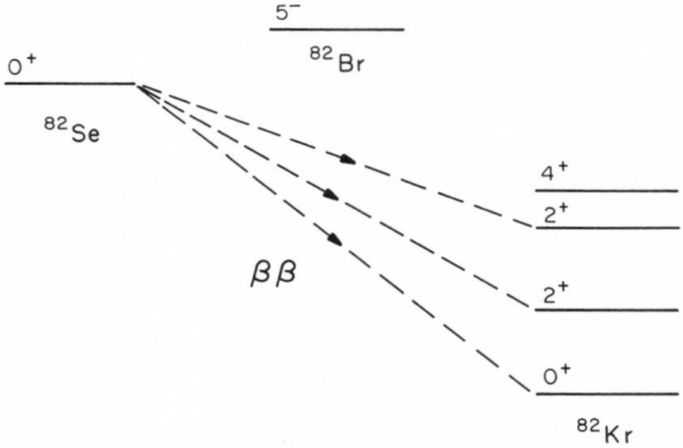

Figure 6.17. Shown is the level diagram for the double beta decay of ^{82}Se. Note that decay to various levels of ^{82}Kr is possible but that ordinary beta decay to ^{82}Br is kinematically forbidden.

Such Majorana masses arise naturally in grand unified theories (GUTS) which attempt to merge the strong and electroweak interactions [29]. This is because in such models the neutrino is placed together with a charged lepton and with quarks of various charges into a multiplet of some unifying group. Because these other particles have mass, we would also expect the neutrino to have a nonzero mass. In the so-called seesaw mechanism, which explains the lightness of the ordinary neutrino states, a state that begins as a four-component Dirac neutrino splits into a pair of two-component Majorana neutrinos—a light one, ν, and a heavy sibling, N, which is as yet undiscovered. The masses are related by [30]

$$m_\nu m_N \sim m_{q,l}^2, \tag{6.55}$$

where m_q, m_l is a typical quark or lepton mass. Thus if N is very heavy—$m_N \gg m_q$—the ν-state is very light.

This distinction between Majorana and Dirac particles was noted early in "neutrino history," and in fact a test for its Majorana/Dirac character was performed by Davis in 1955 [31]. The idea was to permit anti-neutrinos that are emitted in neutron beta decay to strike a neutron "target." If the neutrino is its own anti-particle, then the inverse reaction to p plus e^- should be permitted:

$$\left.\begin{aligned} n &\to p + e^- + \bar{\nu}_e \\ (\bar{\nu}_e = \nu_e) + n &\to p + e^- \end{aligned}\right\} \quad \text{if Majorana.}$$

Davis et al. did not observe the inverse reaction and concluded that the neutrino does not have a Majorana character. Equivalently, a second test of this idea is provided by a search for the so-called neutrinoless double beta decay process:

$${}^A_ZN \to {}_{Z+2}{}^AN + e^- + e^- \quad (+0_\nu).$$

As shown schematically in Figure 6.18a, the sequential emission of two $(e^-\bar{\nu}_e)$ pairs is expected to occur regardless of the nature of the neutrino, with associated lifetimes calculated to be $>10^{20}$ yr. However, if the neutrino is a Majorana particle, then the same neutrino that was emitted in the first weak interaction can be absorbed in the second, giving rise to the neutrinoless process sketched in Figure 6.18b. Early estimates of typical double beta lifetimes under this scenario range from

$$10^{13}\ \text{yr} \lesssim T_{\beta\beta}^{0\nu} \lesssim 10^{15}\ \text{yr}. \tag{6.56}$$

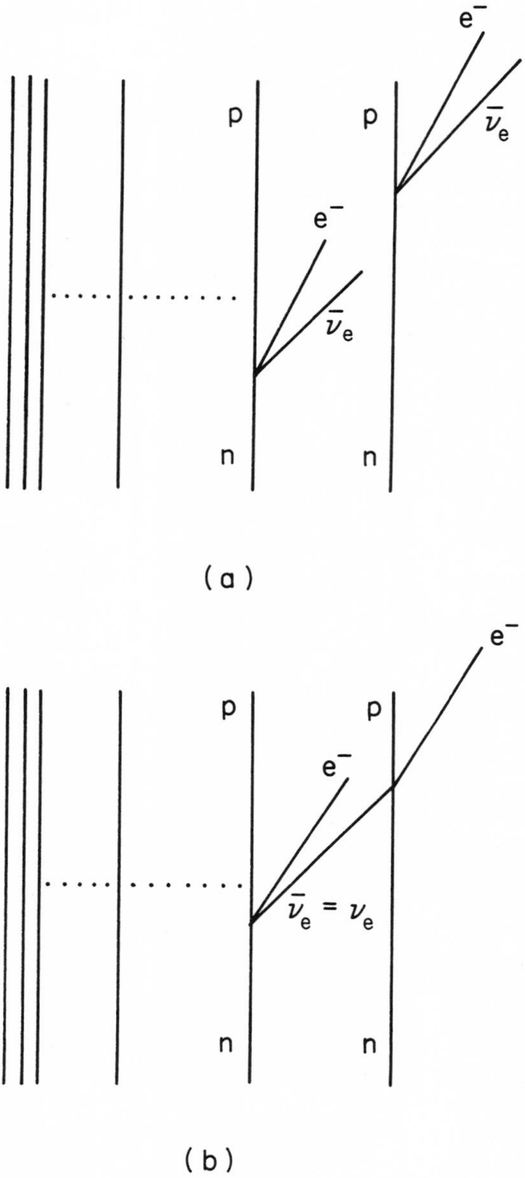

FIGURE 6.18. Two possible mechanisms for double beta decay: (a) In the two-neutrino process, the transition proceeds as a sequential pair of simple neutron beta decays. (b) In the neutrinoless process, the antineutrino emitted in the first decay stimulates the second in the guise of a neutrino.

Comparison with the experimental bounds,

$$T_{\beta\beta}^{\text{expt}} \geq 10^{18}\ \text{yr}, \tag{6.57}$$

again led physicists to abandon the idea of a Majorana neutrino and to propose the idea of lepton number conservation whereby we associate a conserved additive quantum number with each lepton:

$$\begin{array}{ll} \begin{pmatrix} e^- \\ \nu_e \end{pmatrix} L_e = +1 & \begin{pmatrix} e^+ \\ \bar{\nu}_e \end{pmatrix} L_e = -1 \\ \begin{pmatrix} \mu^- \\ \nu_\mu \end{pmatrix} L_\mu = +1 & \begin{pmatrix} \mu^+ \\ \bar{\nu}_\mu \end{pmatrix} L_\mu = -1. \end{array} \tag{6.58}$$

Conservation of lepton number requires that in any process, $A + B \to C + D$, we must require

$$L_e(A + B) = L_e(C + D)$$
$$L_\mu(A + B) = L_\mu(C + D).$$

A Majorana neutrino, which leads to neutrinoless double beta decay, clearly violates the idea of lepton number conservation, since

$${}^{A}_{Z}N \to {}^{\ \ \ A}_{Z+2}N + e^- + e^-$$
$$L_e = 0 \qquad L_e = +2.$$

Thus the absence of double beta zero neutrino processes at the expected rate and the null result established in the Davis experiment led to the idea of this conservation law and to the idea that the neutrino is a Dirac particle. As emphasized earlier, a Majorana particle cannot carry a definite nonzero quantum number.

However, we now know that the situation is not as clear as argued above. The point is that the "standard" model utilizes only left-handed neutrinos,

$$J_\mu{}^w \sim \bar{\psi}_e \gamma_\mu (1 \dotplus \gamma_5) \psi_{\nu_e} + \cdots. \tag{6.59}$$

Then in the Davis experiment, even if the neutrino is a Majorana particle, it has the wrong helicity to initiate the inverse reaction:

$$n \to p + e^- + \bar{\nu}_e(\text{right-handed}),$$

but

$$\nu_e(\text{left-handed}) + n \to p + e^-.$$

Likewise, the neutrinoless double beta decay, $nn \to pp + e^-e^-$, is forbidden even if the neutrino has a Majorana character. Note that these statements are valid only for a massless neutrino, since only in this case is its helicity definite, as noted in chapter 2. (Mathematically this comes about since the propagator of a massless neutrino involves only k_μ, but

$$(1 + \gamma_5)\not{k}(1 + \gamma_5) = 0.) \tag{6.60}$$

Strictly speaking then, we see that the existence of a nonzero neutrinoless double beta decay reaction requires that

(1) the neutrino is a Majorana particle

and

either (2a) $m_\nu \neq 0$ $[(1 + \gamma_5)(\not{k} + m_\nu)(1 + \gamma_5) = 2m_\nu(1 + \gamma_5) \neq 0]$ (6.61)

or

(2b) right-handed currents

$$[\gamma_\mu(1 - \gamma_5)\not{k}\gamma_\nu(1 + \gamma_5) = 2\gamma_\mu\not{k}\gamma_\nu(1 + \gamma_5) \neq 0]. \tag{6.62}$$

(Kayser et al. recently noted that if one assumes the weak interactions to be described by a gauge theory, then even if there exist right-handed currents it is also necessary that at least some of the neutrinos are massive [32]. Otherwise the sum of neutrino contributions must vanish. I shall continue with a purely phenomenological analysis, however.)

More exotic possibilities, such as the existence of a doubly charged Higgs particle, leading to the diagram of Figure 6.19, have also been raised but were shown to be unlikely [33]. Thus we shall concentrate on the two scenarios raised above.

The double beta decay process is often analyzed in terms of the phenomenological weak Hamiltonian,

$$\begin{aligned} H_w \sim \frac{G}{\sqrt{2}} \{&\bar{\psi}_e\gamma_\mu(1 + \gamma_5)\psi_{\nu_e}\bar{u}\gamma^\mu[(1 + \gamma_5) + \eta_{\mathrm{LR}}(1 - \gamma_5)]d \\ &+ \bar{\psi}_e\gamma_\mu(1 - \gamma_5)\psi_{\nu_e}\bar{u}\gamma^\mu[\eta_{\mathrm{RR}}(1 - \gamma_5) + \eta_{\mathrm{RL}}(1 + \gamma_5)]d\} + \text{h.c.}, \end{aligned} \tag{6.63}$$

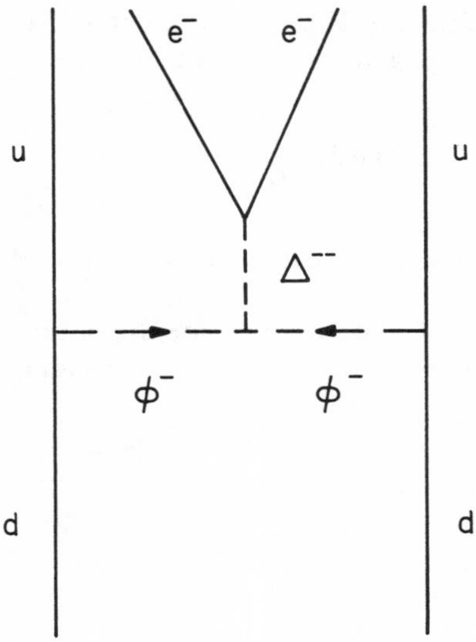

FIGURE 6.19. A highly unlikely (but possible) mechanism for neutrinoless double beta decay. Here ϕ^- is a usual Higgs boson, while Δ^{--} is a doubly charged (exotic) Higgs.

where the (presumably small) couplings $\eta_{\rm LR}$, $\eta_{\rm RL}$, $\eta_{\rm RR}$ vanish in the standard picture. It is then straightforward to evaluate the neutrinoless double beta decay amplitude in impulse approximation, for which we find (assuming $\eta_{ij} = 0$ but $m_\nu \neq 0$) [34]

$$\Gamma_{0\nu} \sim F_0\left(\frac{Q}{m_e}\right) |g_A{}^2 \tilde{M}_{GT} - g_\nu{}^2 \tilde{M}_F|^2 \mathscr{F}^2(Z) m_e{}^7 \frac{\langle m_\nu \rangle^2}{m_e{}^2}, \tag{6.64}$$

where $\mathscr{F}(Z)$ represents a Fermi function,

$$F_0(x) = x\left(1 + 2x + \frac{4}{3}x^2 + \frac{1}{3}x^3 + \frac{1}{30}x^4\right) \tag{6.65}$$

is a kinematic factor and

$$\begin{aligned} \tilde{M}_F &= \left\langle f \left| \frac{1}{2} \sum_{i,j} \tau_i{}^+ \tau_j{}^+ g(r_{ij}) \right| i \right\rangle \\ \tilde{M}_{GT} &= \left\langle f \left| \frac{1}{2} \sum_{i,j} \tau_i{}^+ \tau_j{}^+ g(r_{ij}) \boldsymbol{\sigma}_i \cdot \boldsymbol{\sigma}_j \right| i \right\rangle \end{aligned} \tag{6.66}$$

are nuclear matrix elements. Here $g(r_{ij})$ is the radial dependence associated with the neutrino propagator and has the approximate Yukawa form,

$$g(r_{ij}) \sim \frac{1}{r_{ij}} \exp(-m_\nu r_{ij}). \tag{6.67}$$

A similar form of eq. (6.64) results if $m_\nu = 0$ but $\eta_{ij} \neq 0$. However, I shall not discuss it here.

Likewise, we can calculate the rate for the double beta decay process to occur with the emission of two neutrinos, yielding [34]

$$\Gamma_{2\nu} \sim F_2\left(\frac{Q}{m_e}\right)|g_A{}^2 M_{GT} - g_V{}^2 M_F|^2 \left(\frac{\mathscr{F}(Z)}{E_i - \langle E_n\rangle - \frac{1}{2}E_0}\right)^2 m_e^{11}, \tag{6.68}$$

where

$$F_2(x) = x^7\left(1 + \frac{1}{2}x + \frac{1}{9}x^2 + \frac{1}{90}x^3 + \frac{1}{1980}x^4\right) \tag{6.69}$$

is a kinematic factor, and

$$\begin{aligned} M_F &= \left\langle f \left| \frac{1}{2}\sum_{ij} \tau_i{}^+ \tau_j{}^+ \right| i \right\rangle \\ M_{GT} &= \left\langle f \left| \frac{1}{2}\sum_{ij} \tau_i{}^+ \tau_j{}^+ \boldsymbol{\sigma}_i \cdot \boldsymbol{\sigma}_j \right| i \right\rangle \end{aligned} \tag{6.70}$$

are nuclear matrix elements. Here we have made the so-called closure approximation whereby a sum over all intermediate states is performed by use of an average excitation energy $\langle E_n\rangle$.

We can now confront experimental results with theoretical expectations. The experimental measurements are performed via three quite different techniques. One involves direct laboratory measurement. Moe and Lowenthal originally utilized a cloud chamber to look for the double beta decay of ^{82}Se [35], finding

$$^{82}\mathrm{Se}: T_{1/2}^{2\nu} = (1.0 \pm 0.4) \times 10^{19}\ \mathrm{yr} \tag{6.71}$$

and quoted the number as a 2ν rate since the energy spectrum (both summed and singles) was consistent with this hypothesis. However, these investigators subsequently concluded that the experiment was plagued by certain radioactive backgrounds (with a lifetime of 10^{20} years a mole of ^{82}Se will undergo only one disintegration per hour so that backgrounds must be kept to an absolute minimum), and they redesigned their experiment to involve a time-projection chamber, wherein the paths of the electrons can be followed as they drift under the influence of parallel electric and magnetic fields. Recently they have reported, using this method, the first direct observation of a double beta decay process, with a life-time [36]

$$^{82}\text{Se}: T_{1/2}^{2\nu} \simeq \left(1.1 \begin{matrix} +0.8 \\ -0.3 \end{matrix}\right) \times 10^{20}\ \text{yr}. \tag{6.72}$$

A second approach involves geochemical techniques wherein we infer that a decay has taken place by analyzing an ore sample of known age for the relative isotopic abundances of key elements. Of course, we must know both the age of the ore and that both parent and daughter concentrations have not changed over this (very long) lifetime, something we cannot know for sure. Nevertheless, a number of such measurements have been performed, and typical results are given below:

$$\begin{aligned} &^{130}\text{Te} \to {}^{130}\text{Xe} && T_{1/2} \sim 2.6 \times 10^{21}\ \text{yr} \quad [37] \\ & && \phantom{T_{1/2}} \sim 1.0 \times 10^{21}\ \text{yr} \quad [38] \\ &^{82}\text{Se} \to {}^{82}\text{Kr} && T_{1/2} \sim 1.5 \times 10^{20}\ \text{yr} \quad [39] \end{aligned} \tag{6.73a}$$

as well as the ratio

$$R = \frac{T_{1/2}(^{128}\text{Te})}{T_{1/2}(^{130}\text{Te})} = \begin{matrix} (1.57 \pm 0.10) \times 10^3 \quad [40] & \text{Missouri} \\ > 3.04 \times 10^3. \quad [41] & \text{Heidelberg} \end{matrix} \tag{6.73b}$$

(Note that these two measurements of R are in disagreement. This analysis of Te ore for the presence of residual Xe isotopes is especially challenging for ^{128}Xe, which, because of the exceptionally long half-life, involves only trace amounts of this nuclide over that expected from the atmospheric abundance. Although at present some investigators are inclined to accept the Heidelberg number because of superior statistics, the Missouri group has not capitulated.)

The third type of probe for a double beta decay process is radiochemical, wherein a sample is analyzed within which, in the absence of the double beta decay reaction, one would expect no concentration of a particular daughter nucleus. Examples here include:

$$\begin{array}{ll} {}^{238}\mathrm{U} \rightarrow {}^{238}\mathrm{Pu} & T_{1/2} > 6 \times 10^{18}\ \mathrm{yr.}\ [42] \\ {}^{232}\mathrm{Th} \rightarrow {}^{232}\mathrm{U} & \end{array} \tag{6.74}$$

Now, what does theory have to say? Results of various calculations are summarized in Table 6.1. Obviously, agreement is reasonably satisfactory, considering the complex nature of the nuclear calculations, except in the case of ^{130}Te. However, this difficulty of the nuclear matrix element calculations also suggests that not too much worry be given this discrepancy—indeed, recent investigations suggest that expansion of the shell model basis may provide a resolution.

A mystery possibly remains in the ratio ^{128}Te/^{130}Te, however. Although the decay amplitudes are expected to be nearly the same for these two decays, the lifetimes are quite different due to the very different Q-values for the transitions. Assuming purely two-neutrino decay and identical

TABLE 6.1
CALCULATED VS. EXPERIMENTAL $\beta\beta\nu\nu$ DECAY RATES FOR THREE TRANSITIONS

$\beta\beta$ Reaction	Calculated 2ν Rate (yr)		Expt. (yr)
^{76}Ge → ^{76}Se	1.3×10^{21}	a	$> 2.0 \times 10^{19}$
	4.2×10^{20}	b	
	2.2×10^{20}	c	
^{82}Se → ^{82}Kr	1.2×10^{20}	a	1.5×10^{20} [36, 39]
	0.3×10^{20}	b	
	0.2×10^{20}	c	
^{130}Te → ^{130}Xe	2.2×10^{20}	a	2×10^{21} [37, 38]
	1.7×10^{19}	b	
	1.2×10^{20}	c	

The calculations were done by (a) J. Engel, P. Vogel, and M. R. Zirnbauer, Phys. Rev. *C37*, 731 (1988); (b) W. C. Haxton and G. J. Stephenson, Jr., Prog. Part. Nucl. Phys. *12*, 409 (1984); and (c) H. V. Klapdor and K. Grotz, Nucl. Phys. A460, 395 (1986). The ^{76}Ge limit is given by E. Bellotti et al., Nuovo Cim. *95A*, 1(1986).

matrix elements, we expect

$$\frac{T^{2\nu}_{1/2}(^{128}\mathrm{Te})}{T^{2\nu}_{1/2}(^{130}\mathrm{Te})} = 5130. \tag{6.75}$$

On the other hand, in the case of pure neutrinoless decay we find

$$R \equiv \frac{T^{0\nu}_{1/2}(^{128}\mathrm{Te})}{T^{0\nu}_{1/2}(^{130}\mathrm{Te})} = \begin{cases} 25 & m_\nu \neq 0 \\ 116 & \eta_{\mathrm{RL}} = 0 \end{cases} \tag{6.76}$$

in the case of nonvanishing neutrino mass and right-handed currents, respectively. An attractive feature of this ratio is that it is unaffected by uncertainties in the age of the ore sample and/or possible loss due to diffusion of the Xe by-product.

Comparison with experiment yields no clarification at this time. A Heidelberg measurement [41],

$$R > 3045 \text{ at } 95\% \text{ C.L.}, \tag{6.77a}$$

is obviously consistent with the 2ν-only scenario or with a small 0ν admixture. However, the earlier Missouri number [40],

$$R = 1570 \pm 100, \tag{6.77b}$$

requires the existence of both a two-neutrino as well as a substantial neutrinoless component. A critical ingredient in this analysis is the presumed identity of the ^{128}Te and ^{130}Te matrix elements. Although this is asserted by shell model considerations, the inability to predict accurately the absolute value of the rate of ^{130}Te decay gives one pause. Further nuclear structure work is required before we can really attribute much significance to any possible discrepancy with the ratio predicted on the basis of simple 2ν considerations.

Recently, interest was aroused in an additional mechanism giving rise to double beta decay—that of majoron emission [43]. The idea here is that since the existence of neutrinoless double beta decay would signal the violation of lepton number conservation, as discussed above, it is possible that this symmetry may be spontaneously broken globally. According to Goldstone's theorem (cf. sec. 4.4) this implies the existence of a massless Goldstone boson that has been dubbed the majoron. It has been suggested that if such a particle were to exist, neutrinoless double beta decay could occur accompanied by the majoron. This possibility was raised recently in an experiment involving the decay $^{76}Ge \to {}^{76}Se$ via the double beta process. This is a particularly nice transition to study since Ge detectors are known for their outstanding ($\sim 0.1\%$) resolution and since naturally occurring germanium contains 7.8% of the $A = 76$ isotope. The three possible double beta transitions can then in principle be distinguished:

(1) $^{76}Ge \to {}^{76}Se + 2e^-$; two-body phase space; the energy deposited by the e^- pair should have a strong peak at 2.04 MeV;

(2) $^{76}Ge \to {}^{76}Se + 2e^- +$ majoron; three-body phase space; the energy spectrum has a peak at 1.55 MeV;

(3) $^{76}Ge \to {}^{76}Se + 2e^- + 2\nu$; four-body phase space; the energy spectrum has a broad peak at 0.65 MeV.

Early in 1987 it was reported that just such a bump at ~ 1.55 MeV had been observed, leading to speculation about the existence of a majoron [44]. However, soon thereafter a Santa Barbara/LBL collaboration reported a null signal in a comparable experiment which had lower backgrounds and about an order of magnitude more sensitivity [45]. Thus at the present time there is no evidence for the existence of the majoron.

The latter experiment was also able to set a very stringent limit on the absence of the 0ν mode:

$$T_{1/2}^{0\nu}(^{76}\mathrm{Ge}; 0^+ - 0^+) > 5 \times 10^{23}\ \mathrm{yr}, \qquad (6.78)$$

from which we find a very small upper bound on a possible majorana neutrino mass [46]

$$\langle m_\nu \rangle \gtrsim 1.5\ \mathrm{eV}. \qquad (6.79)$$

Lower bounds on neutrinoless double beta lifetimes have been determined for other systems also, e.g. [47],

$$T_{1/2}^{0\nu}(^{82}\mathrm{Se}) > 3.1 \times 10^{21}\ \mathrm{yr}. \qquad (6.80)$$

However, the neutrino mass bounds arising thereby—$\langle m_\nu \rangle < 15$ eV—are much less sensitive than found via ^{76}Ge.

Clearly much additional work—both experimental and theoretical—is necessary before the double beta decay process and the attendant physics can be said to be understood.

6.5 SOLAR NEUTRINOS

I close this chapter on neutrinos by reviewing the solar neutrino problem, whose origin lies in the experimental attempt to confirm the idea that the sun's energy output is due to nuclear fusion. The fusion hypothesis was put forth nearly fifty years ago, but to this day there has been no *direct* confirmation of its validity—no other hypothesis can account for the essentially steady solar luminosity over a period of five billion years.

The basic energy conversion cycle is from hydrogen to helium and is described in terms of the so-called pp chain, as outlined below [48], [49]:

			Branching ratio (%)	$E_\nu^{\max}$ (MeV)
I	$pp \to de^+\nu_e$		99.75	0.42
	$pe^-p \to d\nu_e$		0.25	1.44
II	$d + p \to {}^3\mathrm{He} + \gamma$		100.00	
III	${}^3\mathrm{He} + {}^3\mathrm{He} \to \alpha + 2p$	or	86.00	
	${}^3\mathrm{He} + \alpha \to {}^7\mathrm{Be} + \gamma$		14.00	
IV	${}^7\mathrm{Be} + e^- \to {}^7\mathrm{Li} + \nu_e$		99.89	0.86 (90%)
	${}^7\mathrm{Li} + p \to \alpha + \alpha$	or		0.36 (10%)
	${}^7\mathrm{Be} + p \to {}^8\mathrm{B} + \gamma$		0.11	
	${}^8\mathrm{B} \to {}^8\mathrm{Be}^* + e^+ + \nu_e$; ${}^8\mathrm{Be}^* \to \alpha + \alpha$			14.06

The rate at which these reactions take place is governed by the standard solar model, wherein the sun is assumed to be in hydrostatic equilibrium, with the gravitational attraction being balanced by the outward pressure due to the burning of the nuclear fuel. Thus, we have

$$\frac{dP}{dr} = -\frac{GM(r)\rho(r)}{r^2}, \tag{6.81}$$

where G is the gravitational constant, $P(r)$ and $\rho(r)$ are the pressure and density at radial distance r from the center of the sun, respectively, and

$$M(r) = \int_0^r dr'\, \rho(r') 4\pi r'^2 \tag{6.82}$$

is the mass contained inside this radius. Also if $L(r)$ is the local luminosity and $\varepsilon(r)$ is the energy production per unit mass, energy balance requires

$$\frac{dL(r)}{dr} = 4\pi r^2 \rho \varepsilon. \tag{6.83}$$

The flow of energy to the surface is controlled by the energy transport equations,

$$\begin{aligned} \frac{dT}{dt} &= -\frac{3}{4\sigma}\frac{\bar{\kappa}}{T^3}\frac{L(r)}{4\pi r^2} && \text{radiative} \\ \frac{dT}{dt} &= \left(1 - \frac{1}{\gamma}\right)\frac{T}{P}\frac{dP}{dt} && \text{convective,} \end{aligned} \tag{6.84}$$

where T is the temperature, σ is the Stefan-Boltzman constant, and $\gamma = C_P/C_V$ is the adiabatic index. The symbol $\bar{\kappa}$, representing the opacity, is calculated via massive computer codes that were developed at Los Alamos and Livermore for other reasons.

Finally, we require the equation of state

$$P = P(\rho, T, f_1, f_2, f_3), \tag{6.85}$$

where f_1, f_2, f_3 are the fractional abundance of hydrogen, helium, and heavier elements. (Of course, the energy production ε and the opacity κ are also dependent on these variables. However, for simplicity of notation I have deleted this feature.) Using the boundary conditions,

$$\begin{aligned} r = 0 \qquad & M(r) = 0 \qquad L(r) = 0 \\ r = R_0 \qquad & M(r) = M_0 = 1.99 \times 10^{30}\ \text{kg} \\ & L(r) = L_0 = 3.90 \times 10^{33}\ \text{erg sec,} \end{aligned} \tag{6.86}$$

these equations can then be integrated numerically to yield the present rate of nuclear reactions and therefore of neutrino production within the sun. (Actually the helium abundance f_2 is not well known since no helium lines are in the visible spectrum. So we use f_2 as a variable, with the value being fixed by demanding that the present luminosity be found when $t = 4.7 \times 10^9$ yr.)

In summary, the standard solar model predicts that neutrinos are emitted from a variety of reactions in the *pp* chain, with predicted fluxes and energies as shown in Table 6.2. Detection of these neutrinos at the predicted flux level would then be direct confirmation of the nuclear fusion hypothesis and would offer a window into the solar interior, which is unavailable by other means.

TABLE 6.2
MAXIMUM ENERGIES AND FLUXES CALCULATED IN THE STANDARD SOLAR MODEL FOR NEUTRINOS EMITTED FROM VARIOUS PARTS OF THE PP CHAIN

Reaction	E_ν^{max} (MeV)	Flux (10^{10}/cm/sec)
$pp \to de^+\nu_e$	0.42	6.1
$^7Be + e^- \to {}^7Li + \nu_e$	0.86 (90%)	4.3×10^{-1}
	0.36 (10%)	
$^8B \to {}^8Be^* + e^+ + \nu_e$	14.06	5.6×10^{-4}

Note that the decay of ^8B to which the Homestake Mine experiment is sensitive constitutes a relatively small fraction of the overall neutrino flux. (W. C. Haxton, in *Intersections Between Particle and Nuclear Physics*, ed. R. Mischke, AIP Conference Proc. No. 123, AIP, New York [1984], p. 1026)

Precisely such an experiment has been underway for many years at the Homestake goldmine in South Dakota by Davis et al. [50]. A (large) tank containing 615 tons of perchloroethylene ($C_2{}^{35}Cl_3{}^{37}Cl_1$) is situated at the 4850-foot level of the mine. The incident solar neutrinos initiate the inverse beta decay reaction, $\nu_e + {}^{37}Cl \to {}^{37}Ar + e^-$. However, the ^{37}Ar is radioactive and can be detected via its subsequent electron capture, $e^- + {}^{37}Ar \to {}^{37}Cl + \nu_e$, which has a 35-day half-life. Over the period since 1970 during which the experiment has been active, it has been determined that 0.47 ± 0.04 ^{37}Ar atoms are produced per day! Since the tank is "swept" for argon only about once a month or so, this means *all* of ten or fifteen atoms! This is obviously no easy task, but there is good reason to suspect that it was done correctly, as evidenced by extensive tests on the recovery efficiency. Results of such experiments are usually quoted in SNU's (standard neutrino units), which are equal to

$$1 \text{ SNU} = 10^{-36} \text{ captures}/{}^{37}\text{Cl atom/sec.} \tag{6.87}$$

Experimentally the capture rate has been found to be [51]

$$R^{exp} = 2.1 \pm 0.3 \text{ SNU}, \tag{6.88}$$

whereas standard solar model calculations by Bahcall give [52]

$$R^{theo} = 8.5 \pm 2.2 \text{ SNU} \tag{6.89}$$

at the 3σ level! Thus the experimental rate is about a factor of three or so smaller than predicted by the standard solar model. This is the "solar" neutrino problem."

At the present time it is not clear what the origin of this discrepancy is, although various possibilities have been suggested. One idea is that some of the experimental cross sections used in predicting reaction rates are incorrect. Thus, for example, ${}^3\text{He} + \alpha \rightarrow {}^7\text{Be} + \gamma$ has been measured experimentally at 100 KeV. However, the cross section is needed at $T \sim 10^7\,^\circ\text{K}$ or 1 KeV for use in the solar codes. Nevertheless, it is felt that this explanation is unlikely and that the required extrapolation is understood.

Another possibility, as raised above, is that the ${}^{37}\text{Ar}$ extraction is incomplete. However, this has been the subject of extensive study and is also thought to be very unlikely.

Other explanations have tried to put the blame on the theoretical calculation, by assuming, for example, that the heavy element abundance in the solar interior is lower than that seen at the surface, or by asserting convective mixing of the core, or by postulating a large central magnetic field density, etc. None of these hypotheses, however, is able to explain the dearth of solar neutrinos without running afoul of some other bit of solar phenomenology [53].

One of the most interesting explanations of the solar neutrino anomaly has to do with the possibility that the origin of the discrepancy lies in the realm of particle physics, that neutrinos mix on their way from their production—in a nuclear process deep within the solar interior—to detection in a huge tank of cleaning fluid in the Homestake mine. Thus, if a neutrino that begins its life as an electron type ν_e oscillates into a ν_μ or ν_τ before reaching the detector, it will be energetically unable to undergo a further charged current reaction and will therefore be effectively sterile thereafter. Originally, the idea was fairly simple. Suppose there are only two neutrino species; then, since the scale of the solar core where such neutrinos are produced is large compared to

$$\frac{E_\nu}{\Delta m^2} \sim 1\ \text{cm} \quad \text{for } \Delta m^2 \sim 10^{-2}\ \text{eV}^2, \tag{6.90}$$

we find from eq. (6.38) that

$$P_{\nu_e} \sim 1 - \sin^2 2\theta \sin^2 \overline{\frac{\Delta m^2 L}{2E_\nu}} \sim 1 - \frac{1}{2}\sin^2 2\theta, \tag{6.91}$$

where the average is over both E_ν and L. Provided we have maximal mixing—$\theta \sim 45°$—we find

$$P_{\nu_e} \simeq \frac{1}{2}, \tag{6.92}$$

which could allow a factor of two reduction in the predicted ^{37}Cl production rate. In a three-neutrino mixing scenario with the mixing angles optimized, we could imagine the rate being reduced by the observed factor of three. However, this hypothesis has always seemed somewhat artificial, requiring the mixing angles to be chosen precisely. If, for example, we take the two-neutrino angle as the Cabibbo angle

$$\theta \sim \theta_c = 15°, \tag{6.93}$$

then

$$P_{\nu_e} \simeq 1 - \frac{1}{2}\sin^2 2\theta_c \simeq 0.88, \tag{6.94}$$

which represents only a 12% reduction.

Recently, however, a new suggestion has generated a good deal of excitement in the field. Mikheyev and Smirnov pointed out that flavor oscillations of solar neutrinos might be greatly enhanced, allowing sensitivity to small mixing angles and neutrino mass differences [54]. The mechanism which they propose depends upon an effective density-dependent contribution to the mass of the electron neutrino generated by charged current scattering of solar electrons, a phenomenon first noted by Wolfenstein [55]. For this reason the mechanism is usually called the MSW effect.

The basic idea is well known from elementary quantum mechanics. Consider a simple two-state system whose energy levels are a function of some parameter—say, an external magnetic field—as shown in Figure 6.20. If we now change the value of this parameter *slowly* compared to a typical oscillation time, the adiabatic theorem guarantees that if we begin, say, on the branch with lowest energy, then we will remain on this branch as the parameter varies. Application to the solar neutrino problem is provided by the feature that the neutrino mass matrix is a function of the solar density.

The mechanism by which the density affects the neutrino mass is that while electron type neutrinos can scatter elastically from solar electrons by means both of the charged *and* neutral weak current (Figure 6.21a),

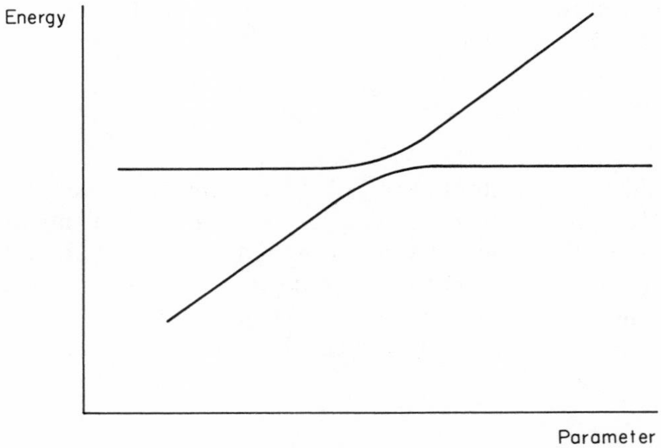

FIGURE 6.20. A generic energy level plot for a two-state system, whose energy is dependent on an external parameter. If the parameter is changed adiabatically, the system will stay in the same (parameter-dependent) level in which it begins.

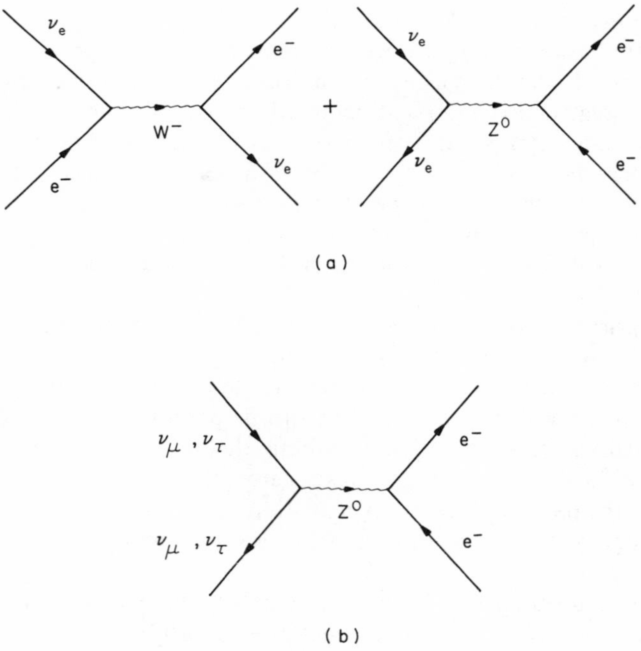

FIGURE 6.21. While both neutral *and* charged current mechanisms are possible for $\nu_e e^-$ scattering (a), only the neutral current process is possible for $\nu_\mu e^-$ or $\nu_\tau e^-$.

only the neutral current option is available to ν_μ or ν_τ (Figure 6.21b). This difference implies that the interaction energy of the electron neutrino will be slightly different from that of neutrinos of other generations, and the size of the effect will be dependent on the local electron density. Writing the neutrino wave function in two-component form as

$$|\psi_\nu(t)\rangle = \begin{pmatrix} a_1(t) \\ a_2(t) \end{pmatrix}, \tag{6.95}$$

where

$$|\nu_1\rangle = \begin{pmatrix} 1 \\ 0 \end{pmatrix} \quad \text{and} \quad |\nu_2\rangle = \begin{pmatrix} 0 \\ 1 \end{pmatrix} \tag{6.96}$$

are the mass eigenstates, the equation of motion becomes

$$i\frac{d}{dt}\begin{pmatrix} a_1(t) \\ a_2(t) \end{pmatrix} = \begin{pmatrix} \dfrac{m_1{}^2}{2p} + \sqrt{2}G\rho\cos^2\theta, & \sqrt{2}G\rho\cos\theta\sin\theta \\ \sqrt{2}G\rho\cos\theta\sin\theta, & \dfrac{m_2{}^2}{2p} + \sqrt{2}G\rho\sin^2\theta \end{pmatrix}\begin{pmatrix} a_1(t) \\ a_2(t) \end{pmatrix} \tag{6.97}$$

where $\rho(t)$ is the local electron density. If the density were constant, we could solve for the energy eigenstates, yielding

$$\begin{aligned} |\nu_e\rangle &= \cos\chi(\rho)|\nu_1(\rho)\rangle + \sin\chi(\rho)|\nu_2(\rho)\rangle \\ |\nu_\tau\rangle &= -\sin\chi(\rho)|\nu_1(\rho)\rangle + \cos\chi(\rho)|\nu(\rho)\rangle \end{aligned} \tag{6.98}$$

with

$$\tan 2\chi(\rho) = \frac{\sin 2\theta}{\cos 2\theta + \dfrac{2\sqrt{2}G\rho p}{m_1{}^2 - m_2{}^2}}. \tag{6.99}$$

Obviously at zero density the mixing angle is equal to θ and the eigenstates are the simple mass eigenstates. On the other hand, at very large density, we find, assuming $m_2 > m_1$, that

$$\lim_{\rho\to\infty} \chi(\rho) = \frac{\pi}{2}. \tag{6.100}$$

The eigenenergies are shown in this case as a function of the density in Figure 6.22. The critical density, corresponding to that value where

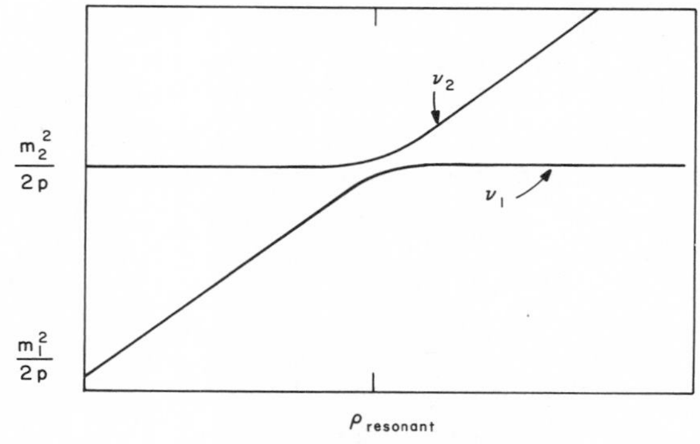

FIGURE 6.22. Plotted are the eigenvalues of eq. (6.97) for the case $m_2 > m_1$ and $\sin\theta \ll 1$ as a function of electron density.

$\chi(\rho_{\rm crit}) = \dfrac{\pi}{4}$ is given by

$$\rho_{\rm crit} = 4.0 \times 10^{24}\ {\rm cm}^{-3} \times \frac{1\ {\rm MeV}/c}{p} \times \frac{m_2{}^2 - m_1{}^2}{10^{-6}\ {\rm eV}^2}. \tag{6.101}$$

Since

$$|\nu_2(\rho)\rangle = \sin\chi(\rho)|\nu_e\rangle + \cos\chi(\rho)|\nu_\tau\rangle, \tag{6.102}$$

we see that a neutrino produced as ν_e at a density greater than $\rho_{\rm crit}$, where $\chi(\rho) \approx \dfrac{\pi}{2}$, is essentially in the eigenstate $|\nu_2(\rho)\rangle$. Provided the passage from the production region, where $\rho > \rho_{\rm crit}$, to the solar exterior, where $\rho = 0$, is adiabatic, then when the neutrino reaches the surface its state will be

$$\sin\theta|\nu_e\rangle + \cos\theta|\nu_\tau\rangle, \tag{6.103}$$

which is predominantly $|\nu_\tau\rangle$ for small mixing angles. Thus the deficit of solar neutrinos arises in a "natural" fashion in this scenario and does not require any "special" value of the mixing angle. Careful study of the solution of these mixing equations reveal a large region of $\theta,\delta m^2$ space which is consistent with the ^{37}Cl experiment, as shown in Figure 6.23. For nearly any mixing angle we find a possible solution. Also we note that this model is sensitive to very small values of neutrino mass differences,

$$10^{-4}\ {\rm eV}^2 \geq \delta m^2 \geq 10^{-8}\ {\rm eV}^2, \tag{6.104}$$

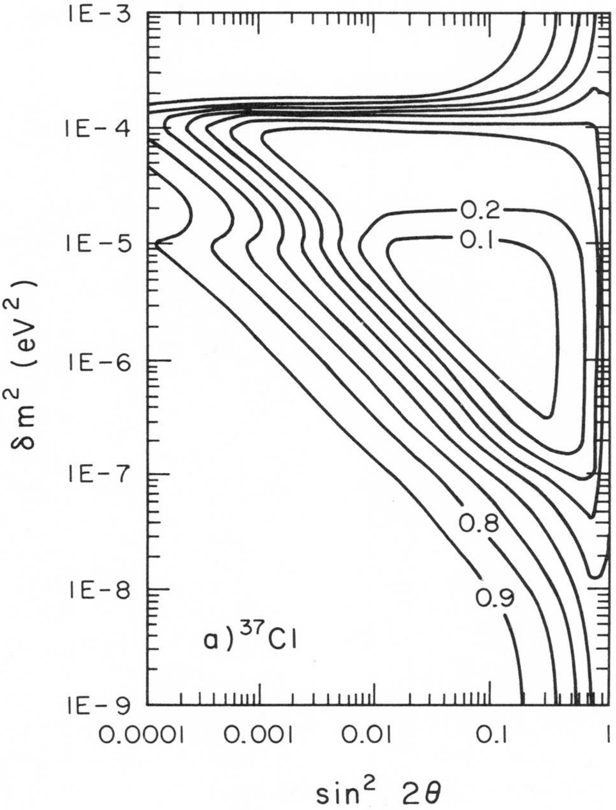

FIGURE 6.23. Plotted is the ^{37}Cl capture rate in fractions of the standard model value [56]. In these units the experimental value found by Davis et al. [51] is $0.15 \lesssim f \lesssim 0.5$.

which are far beyond the capability of any terrestrial experiment to detect. As mentioned above, such masses might be expected to arise, for example, in various grand unified models wherein very heavy gauge bosons—$M_B \gtrsim 10^{14}$ GeV—can lead, via the seesaw process, to tiny neutrino masses. For all these reasons the MSW effect has gained a great deal of attention as a possible explanation of the solar neutrino anomaly.

In order to really address this problem and find out which (if any) of these proposed resolutions is correct, it is essential to obtain additional experimental information. We observe that the ^{37}Cl capture reaction is not primarily to the ^{37}Ar ground state, which has a 814 KeV threshold and log ft $\sim$ 5, but rather to the analog level at 4.98 MeV with log ft $\sim$ 3. Because of this, the ^{37}Cl detector is sensitive mainly to the very high energy neutrinos arising predominantly from the decay of ^8B, which represents only a tiny percentage ($<0.1\%$) of the *pp* neutrinos that initiated the fusion process. This dominance of these high energy ^8B neutrinos can be seen by breaking the predicted capture rate down into its various components, yielding [56]

$$R^{\text{theo}} = 6.1 \text{ SNU}(^8\text{B}) + 1.2 \text{ SNU}(^7\text{Be}) + \ldots = 7.8 \text{ SNU}. \quad (6.105)$$

From this point of view it is profitable to find a reaction that is sensitive to lower-energy neutrinos coming from the primary *pp* reaction

$$pp \to d + e^+ + \nu_e \qquad E_\nu^{\max} = 0.42 \text{ MeV}.$$

Such a detector has been proposed, using the reaction

$$\nu_e + {}^{71}\text{Ga} \to {}^{71}\text{Ge} + e^-.$$

Although there exists some uncertainty in this case about the cross sections to the $\frac{5}{2}^-$ (175 KeV) and $\frac{3}{2}^-$ (500 KeV) excited states, the dominant capture is to the ground state in ^{71}Ge, which has a threshold of only 233 KeV. Thus the primary sensitivity in this case should be to the *pp* neutrinos, and indeed a calculation over the spectrum yields [56]

$$R^{\text{theo}} = 70 \text{ SNU}(pp) + 33 \text{ SNU}(^7\text{Be}) + 15(^8\text{B}) + \ldots = 128 \text{ SNU}.$$

A gallium experiment would then offer a clear view into the primary $pp \to de^+\nu_e$ reaction that initiates the solar burning chain. Such experiments are currently underway in Western Europe and in the Soviet Union. In the former case the experiment will employ 30 metric tons of Ga in the form of a $GeCl_3$–HCl solution. The ^{71}Ge byproduct has a half-life of 11.4d and will be extracted as $GeCl_4$ by means of a gas purge. The $GeCl_4$ is then converted to GeH_4, which is counted in small proportional counters. The Russian experiment, on the other hand, utilizes fifty tons of metallic gallium and a different and somewhat less well understood chemistry. In both cases it will be several years before results are obtained. Nevertheless the measured ^{71}Ge production rate will provide crucial input to the solar neutrino puzzle since the primary neutrino source is the beginning *pp* process. If the MSW effect is the reason for the paucity of ^{37}Cl events, then the measured ^{71}Ga event rate will be sensitive to a very different region of $\theta,\delta m$ space than for ^{37}Cl and should enable us to pin down the mixing angle and mass difference rather precisely, as shown in Figure 6.24.

For completeness we should mention that a number of additional solar neutrino detection schemes have been proposed, though none have yet achieved funding. Perhaps the most promising is the suggestion by a joint U.S./Canadian collaboration to place a large tank of heavy water (D_2O) in a nickel mine in Sudbury, Ontario. This proposal is unique because it can directly test the neutrino mixing hypothesis because it is sensitive to both charged *and* neutral current neutrino processes. The former can be seen via the reaction

$$\nu_e + d \to p + p + e^- \quad (E_\nu^{\min} = 1.44 \text{ MeV}),$$

while the latter would lead to the breakup channel,

$$\nu + d \to p + n + \nu \quad (E_\nu^{\min} = 2.2 \text{ MeV}).$$

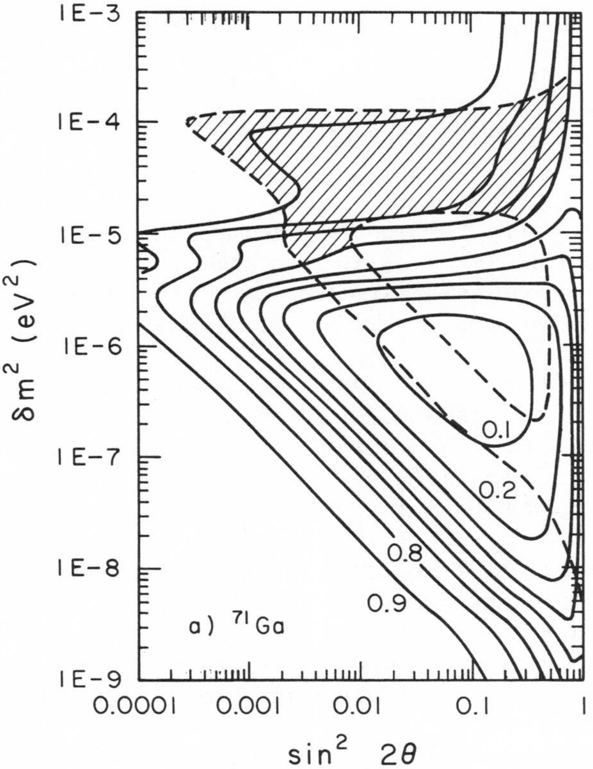

FIGURE 6.24. Plotted is the ^{71}Ge capture rate in fractions of the standard model value [56]. Here the dashed lines represent the "allowed" region from the ^{37}Cl experiment.

Only electron neutrinos can participate in the charged-current process while any neutrino species can lead to the deuterium breakup. Thus comparison of the experimental fluxes would lead to a direct measurement of the $\nu_e/\sum_i \nu_i$ ratio.

This is an exciting time for solar neutrino physics: for the first time in history, we are able to look almost directly into the solar interior and study the basic mechanism of energy production that makes life on Earth possible.

References

[1] This assertion is, of course, only true in lowest order. The neutrino may have a charge radius, for example, and there have been recent speculations about the implications of a possible magnetic moment.

[2] A. Sandage and G. A. Tammann, Ap. J. *196*, 313 (1975), and *197*, 265 (1975).

[3] H. C. Ohanian, *Gravitation and Spacetime*, Norton, New York (1976), Ch. 10.6; S. Weinberg, *Gravitation and Cosmology*, Wiley, New York (1972), Ch. 15.2.

[4] P.J.E. Peebles, *Physical Cosmology*, Princeton Univ. Press, Princeton, N.J. (1974), Ch. 4.

[5] See, e.g., *Nearly Normal Galaxies*, ed. S. M. Faber, Springer-Verlag, New York (1987), and references therein.

[6] For a recent review, see *Dark Matter in the Universe*, ed. J. Knapp and J. Kormendy, Reidel, New York (1986), Proc. of Int. Ast. Union Symp. No. 117.

[7] T. P. Cheng and L. F. Li, *Gauge Theory of Elementary Particle Physics*, Oxford Univ. Press, Oxford (1984), Ch. 11.

[8] D. B. Cline, D. N. Schramm, and G. Steigman, Comm. Nuc. Part. Phys. *17*, 145 (1987).

[9] E. D. Commins, *Weak Interactions*, McGraw-Hill, New York (1973), Ch. 14.

[10] G. Steigman, D. N. Schramm, and J. Gunn, Phys. Lett. *B66*, 502 (1977); J. Yang et al., Ap. J. *281*, 493 (1984).

[11] J. F. Cagaignac et al., Phys. Lett. *B148*, 387 (1984).

[12] L. S. Durkin et al., Phys. Rev. Lett. *61*, 1811 (1988).

[13] See, e.g., *Nuclear Physics*, E. Fermi, Univ. of Chicago Press, Chicago (1962), Ch. 4.

[14] J. J. Simpson, Phys. Rev. Lett. *54*, 1891 (1985), and Phys. Lett. *B174*, 113 (1986).

[15] S. Boris et al., Phys. Lett. *B159*, 217 (1985), and Phys. Rev. Lett. *58*, 2019 (1987).

[16] T. Ohi et al., Phys. Lett. *B160*, 322 (1985); J. Markey and F. Boehm, Phys. Rev. *C32*, 2215 (1985); T. Altzitzoglou et al., Phys. Rev. Lett. *55*, 799 (1985).

[17] W. C. Haxton, Phys, Rev. Lett. *55*, 807 (1985); J. Lindhard and P. G. Hansen, Phys. Rev. Lett. *57*, 965 (1986).

[18] J. F. Wilkerson et al., Phys. Rev. Lett. *58*, 2023 (1987); see also R.G.H. Robertson and D. A. Knapp, to be published in Ann. Rev. Nucl. Part. Sci.

[19] M. Fritschi et al., Phys. Lett. *B173*, 485 (1986).

[20] K. E. Bergkvist, Phys. Lett. *B154*, 224 (1985), and *B159*, 408 (1985).

[21] C. L. Bennett et al., Phys. Rev. *C31*, 197 (1985).

[22] C. Ching and T. Ho, Phys. Rept. *112*, 1 (1984).

[23] R.G.H. Robertson in *Intersections Between Particle and Nuclear Physics*, ed. D. F. Geesaman, AIP Conf. Proc. No. 150, AIP, New York (1986), p. 115.

[24] H. L. Ravn et al., in *Neutrino Mass and Gauge Structure of Weak Interactions*, eds. V. Barger and D. Cline, AIP Conf. Proc. No. 99, AIP, New York (1983), p. 1.

[25] S. Chandrasekhar, Mon. Not. Roy. Ast. Soc. *95*, 207 (1935); L. D. Landau, Phys. Z. Sowjet Union *1*, 285 (1932).

[26] H. Bethe and G. E. Brown, Sci. Am. *262*, 60 (1985); E. Baron et al., Phys. Rev. Lett. *55*, 126 (1985).

[27] K. Hirata et al., Phys. Rev. Lett. *58*, 1490 (1987); R. Bionta et al., Phys. Rev. Lett. *58*, 1494 (1987).

[28] E. W. Kolb, A. Stebbins, and M. S. Turner, Phys. Rev. *D35*, 3598 (1987); S. Bahcall and S. D. Glashow, Nature *326*, 476 (1987); A. Burrows and J. Lattimer, Ap. J. *318*, L63 (1987).

[29] G. G. Ross, *Grand Unified Theories*, Benjamin-Cummings, Reading, Mass. (1985).

[30] T. P. Cheng and L. F. Li, *Gauge Theory of Elementary Particle Physics*, Oxford Univ. Press, Oxford (1984), p. 419; M. Gell-Mann, P. Ramond, and R. Slansky in *Supergravity*, ed. D. Z. Freedman and P. Van Nieuwenhuizen, North-Holland, New York (1979).

[31] R. Davis, Phys. Rev. *97*, 766 (1955).

[32] B. Kayser, in *Searches for New and Exotic Phenomena*, ed. O. Fackler and T. Van Tranh Proc. Seventh Moriond Workshop (1987).

[33] R. N. Mohapatra and J. D. Vergados, Phys. Rev. Lett. *47*, 1713 (1981); L. Wolfenstein, Phys. Rev. *D26*, 2507 (1982); J. Schechter and J.W.F. Valle, Phys. Rev. *D25*, 2951 (1982).

[34] See, e.g., W. C. Haxton and G. J. Stephenson, Jr., Prog. in Part. Nucl. Phys. *12*, 409 (1984). F. T. Avignone and R. L. Brodzinski, Prog. in Part. Nucl. Phys. *21*, 99 (1988).

[35] M. K. Moe and D. D. Lowenthal, Phys. Rev. *C23*, 2186 (1980).
[36] S. R. Elliott, A. A. Hahn, and M. K. Moe, Phys. Rev. Lett. *59*, 2020 (1987).
[37] T. Kirsten, H. Richter, and E. Jessberger, Phys. Rev. Lett. *50*, 474 (1983).
[38] E. W. Hennecke, O. K. Manuel, and D. D. Sabu, Phys. Rev. *C11*, 1378 (1975).
[39] T. Kirsten in *Science Underground*, ed. M. M. Nieto, AIP Conf. Proc. No. 96, AIP, New York (1983), p. 396
[40] E. Hennecke et al, ref. 38.
[41] T. Kirsten et al., Phys. Rev. Lett. *51*, 474 (1983).
[42] C. A. Levine, A. Ghiorso, and G. T. Seaborg, Phys. Rev. *27*, 296 (1950).
[43] Y. Chikasage et al., Phys. Lett. *B98*, 265 (1981); G. B. Gelmini and M. Roncadelli, Phys. Lett. *B99*, 411 (1981); H. M. Georgi et al., Nucl. Phys. *B193*, 297 (1981).
[44] F. T. Avignone et al., Proc. Ann. Meeting of APS Div. Part. & Fields, Salt Lake City, Utah (1987), to be published, New York Times, 14 January 1987; M. Waldrop, Science *235*, 534 (1987).
[45] D. O. Caldwell et al., Phys. Rev. Lett. *59*, 419 (1987); see also P. Fisher et al., Phys. Lett. *B192*, 460 (1987).
[46] T. Kotani in Proc. 12th Int. Conf. on Neutrino Physics and Astrophysics, ed. T. Kitagaki and H. Yuta, World Scientific, Singapore (1986), p. 114.
[47] M. Doi, T. Kotani, and E. Takasugi, Prog. Theor. Phys. Suppl. *83*, 1 (1985); W. C. Haxton and G. J. Stephenson, Jr., ref. 34; K. Grotz and H. V. Klapdor, Phys. Lett. *B153*, 1 (1985).
[48] E. C. Commins, *Weak Interactions*, McGraw-Hill, New York (1973), Ch. 14.2.1; J. N. Bahcall and R. M. May, Ap. J. *155*, 501 (1969); H. A. Bethe and C. L. Critchfield, Phys. Rev. *54*, 248 (1938).
[49] J. N. Bahcall et al., Rev. Mod. Phys. *54*, 767 (1982).
[50] R. Davis et al., Phys. Rev. Lett. *20*, 1205 (1968).
[51] J. K. Rowley et al., in *Solar Neutrinos and Neutrino Astronomy* ed. M. L. Cherey, W. A. Fowler, and K. Lande, AIP Conf. Proc. No. 126, AIP, New York (1985). p. 1.
[52] J. N. Bahcall, Rev. Mod. Phys. *59*, 505 (1987); see also J. Bahcall, *Neutrino Astrophysics*, to be published by Cambridge Univ. Press.
[53] W. C. Haxton, Comm. Nucl. Part. Phys. *16*, 95 (1986).
[54] S. P. Mikheyev and A. Yu Smirnov, Sov. J. of Nucl. Phys. *42*, 913 (1985), and Nuovo Cimento *9*, 17 (1986).
[55] L. Wolfenstein, Phys. Rev. *D17*, 2369 (1978).
[56] W. C. Haxton, Phys. Rev. *D35*, 2352 (1987).

Chapter 7

PROBES OF NUCLEAR PHYSICS

By just exchange one for the other given:
. . . .
There never was a better bargain driven
—PHILIP SIDNEY

7.1 Meson Exchange Currents

Thus far in this presentation I have used weak interactions in the nuclear medium as a laboratory in which to study something basic about the interaction itself or about the fundamental particles—quarks or leptons—that participate in that interaction. It is also possible, however, to use the weak interaction to reveal information about the nucleus, and in this final chapter I will give one such example.

The question I will address is this: Do pions exist inside the nucleus? In elementary nuclear physics courses we learn that at a first level of approximation the nucleus is merely a collection of neutrons and protons, with a mass given by

$$M \approx Zm_p + (A - Z)m_n + \ldots, \tag{7.1}$$

where Z,A are the atomic number and mass number, respectively. At the next level of magnification it is, of course, necessary to include the effects of nuclear binding. In the so-called liquid drop model, this is represented as [1]

$$M - Zm_p - (A - Z)m_n \approx -a_1 A + a_2 A^{2/3} + \ldots, \tag{7.2}$$

where $a_1 \approx 16$ MeV is an average volume binding energy, $a_2 \approx 18$ MeV signifies a surface correction, and the ellipses designate smaller Coulombic, pairing, and other effects. Also, at the simplest level one asscribes the origin of the nuclear binding force as due to the exchange of virtual mesons—π, ρ, ω, etc.—between nucleons. Of these the pion is most important because of its light mass, and one would anticipate then that "smoking gun" evidence of the presence of these pionic degrees of freedom should be relatively easy to find. As we shall see, however, this is not at all the case.

First, we review the nature of the pion exchange interaction. Recall that in electrodynamics the potential outside an electric dipole of strength

$\mathbf{p}_1$ located at position $\mathbf{r}_1$ is given by [2]

$$\phi(\mathbf{r}) = -\mathbf{p}_1 \cdot \nabla \frac{1}{4\pi|\mathbf{r} - \mathbf{r}_1|}, \tag{7.3}$$

leading to an interaction energy between it and a second dipole $\mathbf{p}_2$ at location $\mathbf{r}_2$

$$U = -\mathbf{p}_2 \cdot \mathbf{E}_1(\mathbf{r}_2) = \frac{1}{3}\mathbf{p}_1 \cdot \mathbf{p}_2 \delta^3(\mathbf{r}_1 - \mathbf{r}_2) - \frac{1}{4\pi|\mathbf{r}_1 - \mathbf{r}_2|^3}(3\mathbf{p}_1 \cdot \hat{r}_{12}\mathbf{p}_2 \cdot \hat{r}_{12} - \mathbf{p}_1 \cdot \mathbf{p}_2). \tag{7.4}$$

With this knowledge we can now derive the corresponding expression for the pion-exchange interaction. This is because in the nonrelativistic approximation the pion field outside a nucleon fixed at location $\mathbf{r}_1$ is given by [3]

$$\phi_1{}^a(\mathbf{r}) = \frac{f}{m_\pi} \tau^a \boldsymbol{\sigma} \cdot \nabla \frac{e^{-m_\pi|\mathbf{r}-\mathbf{r}_1|}}{|\mathbf{r} - \mathbf{r}_1|}, \tag{7.5}$$

where a signifies the isotopic spin index of the pion

$$a: \begin{cases} \frac{1}{\sqrt{2}}(1 + i2): \pi^+ \\ 3: \pi^0 \\ \frac{1}{\sqrt{2}}(1 - i2): \pi^- \end{cases} \tag{7.6}$$

and f is related to the usual πNN coupling via

$$f = g_{\pi NN} \frac{m_\pi}{2m_N}. \tag{7.7}$$

Thus the nuclear spin generates a dipolelike pionic field in its vicinity very much as in the electrostatic case. An important difference is the existence of the term $\exp(-m_\pi r)/r$ as opposed to the Coulombic $1/r$ potential. This arises, of course, from the feature that the carrier of the interaction is massive, leading to the familiar Yukawa form,

$$\int \frac{d^3q}{(2\pi)^3} \frac{e^{i\mathbf{q}\cdot\mathbf{r}}}{\mathbf{q}^2 + m_\pi{}^2} = \frac{\exp(-m_\pi r)}{4\pi r}. \tag{7.8}$$

The pion exchange interaction now follows directly in analogy to eq. (7.4):

$$\begin{aligned} U_{\pi NN} &= \frac{f^2}{m_\pi^2} \boldsymbol{\tau}_1 \cdot \boldsymbol{\tau}_2 \boldsymbol{\sigma}_2 \cdot \boldsymbol{\nabla}_2 \boldsymbol{\sigma}_1 \cdot \boldsymbol{\nabla}_1 \frac{e^{-m_\pi|\mathbf{r}_1 - \mathbf{r}_2|}}{|\mathbf{r}_1 - \mathbf{r}_2|} \\ &= \frac{1}{3}\frac{f^2}{4\pi}\left(1 + \frac{3}{m_\pi r_{12}} + \frac{3}{(m_\pi r_{12})^2}\right)\frac{e^{-m_\pi r_{12}}}{r_{12}} \\ &\times (3\boldsymbol{\sigma}_1 \cdot \hat{r}_{12} \boldsymbol{\sigma}_2 \cdot \hat{r}_{12} - \boldsymbol{\sigma}_1 \cdot \boldsymbol{\sigma}_2) \boldsymbol{\tau}_1 \cdot \boldsymbol{\tau}_2 \\ &+ \frac{1}{3} f^2 \left(\frac{e^{-m_\pi r_{12}}}{4\pi r_{12}} - \frac{1}{m_\pi^2} \delta^3(\mathbf{r}_1 - \mathbf{r}_2)\right) \boldsymbol{\sigma}_1 \cdot \boldsymbol{\sigma}_2 \boldsymbol{\tau}_1 \cdot \boldsymbol{\tau}_2. \end{aligned} \tag{7.9}$$

The delta-function component is usually discarded on the grounds that the short-range nucleon-nucleon repulsion (i.e., "hard core") makes it inoperative. The remaining pieces of the interaction potential are certainly present, and abundant evidence exists of this fact. For example,

(1) *Nucleon-nucleon scattering lengths.* Since the pion is so light, the long-range piece of the nucleon-nucleon force should be dominated by the pion exchange contribution. One should be able to see this by comparing experimental and theoretical expressions for the scattering lengths of the high partial waves, for which one has in Born approximation [4]

$$a_{l\pi}^{\text{Born}} = \frac{M}{((2l+1)!!)^2} \int_0^\infty dr\, r^{2l+2} U_\pi(r). \tag{7.10}$$

Realistically this program can only be carried out in the *P*-wave channels since *S*-waves are also affected by short-range effects arising from heavier meson exchanges, while for *D*,*F*,*G*, . . . partial waves the experimental data are not sufficiently reliable to allow a meaningful confrontation between experiment and theory. The *P*-wave results are summarized in Table 7.1. Obviously, agreement is not perfect. On the other hand, *P*-waves are hardly the high partial waves where one would expect π-exchange only to be operative, and yet the pion fingerprint is clearly present in the experimental numbers.

(2) *D-waves and the deuteron.* The presence of the "tensor force" component

$$3\boldsymbol{\sigma}_1 \cdot \hat{r}_{12} \boldsymbol{\sigma}_2 \cdot \hat{r}_{12} - \boldsymbol{\sigma}_1 \cdot \boldsymbol{\sigma}_2 \tag{7.11}$$

leads to the prediction of a small but substantial *D*-wave amplitude in the predominantly *S*-wave deuteron wave function. Since the deuteron is quite loosely bound, we expect pion exchange to play a critical role. Thus writing

$$\psi = \frac{1}{r} u(r) \mathscr{Y}_{101} + \frac{1}{r} w(r) \mathscr{Y}_{121}, \tag{7.12}$$

TABLE 7.1
COMPARISON OF THEORETICAL AND EXPERIMENTAL VALUES OF *P*-WAVE *NN-SCATTERING VOLUMES*

P-Wave Scattering Channel	Theory (fm^3) (Pion Exchange)	Experiment (fm^3)
$a(^1P_1)$	3.07	~ 3
$a(^3P_0)$	-3.28	-2.7 ± 0.3
$a(^3P_1)$	2.19	2.0 ± 0.1
$a(^3P_2)$	0	-0.3 ± 0.03

Comparison between *P*-wave nucleon-nucleon scattering volumes as calculated from $U_\pi(r)$ in Born approximation and as measured experimentally [4].

where $u(r)$, $w(r)$ represent *S*,*D*-wave components, respectively, we define the D/S ratio by evaluating these radial amplitudes in the asymptotic regime,

$$\eta_{DS} \equiv \lim_{r \to \infty} \frac{w(r)}{u(r)}. \tag{7.13}$$

Comparison then between the experimental and theoretical values of η_{DS} [5]

$$\begin{aligned} \eta_{DS}(\text{theory}) &= 0.02762 \\ \eta_{DS}(\text{expt.}) &= 0.0271(6) \end{aligned} \tag{7.14}$$

reveals remarkable agreement, confirming the presence of the tensor force predicted in eq. (7.9).

This can also be observed by looking at the deuteron quadrupole moment, which is given theoretically as [5]

$$Q = \frac{1}{\sqrt{50}} \int_0^\infty dr\, r^2 u(r) w(r) - \frac{1}{20} \int_0^\infty dr\, r^2 w^2(r). \tag{7.15}$$

The *D*-wave amplitude is seen to play a critical role, and once again experimental measurements provide a strong confirmation of theoretical predictions [5]:

$$\begin{aligned} Q(\text{theory}) &= 0.2785 \text{ fm}^2 \\ Q(\text{expt.}) &= 0.2859(3) \text{ fm}. \end{aligned} \tag{7.16}$$

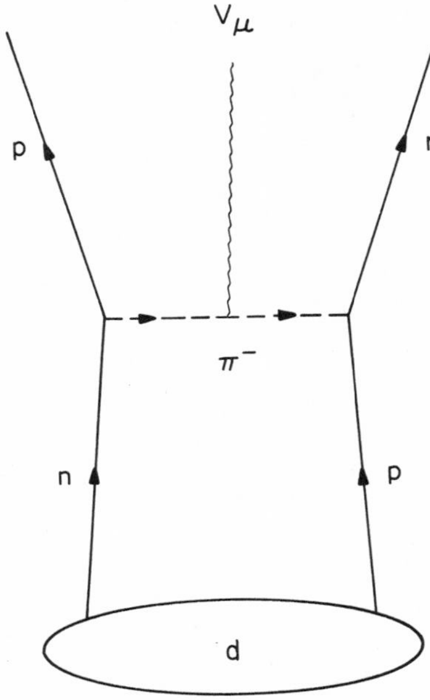

FIGURE 7.1. Shown is a meson exchange diagram relevant for deuteron photo-(electro-) disintegration.

From this analysis we may say that the existence of the pion exchange component in the nucleon-nucleon interaction is substantially confirmed. However, that is not of concern here. Rather, we want to know whether evidence exists for the presence of pionic degrees of freedom in the nucleus itself—not just in the tail of the nucleon-nucleon interaction.

Two important pieces of evidence are often cited in this regard:

(1) *Electrodisintegration of the deuteron.* A standard—textbook—answer to the question of existence of pionic degrees of freedom comes from the electrodisintegration of the deuteron, $e^- + d \rightarrow n + p + e^-$, the theoretical analysis of which involves transition to the $^1S_0$ np state and utilizes contributions from both one-body (impulse approximation) and two-body (meson exchange) currents [6]. The latter are shown in Figure 7.1 and have the form

$$\mathbf{J} \sim \mathbf{q} \times [h^{(1)}(r)(\tau_1 - \tau_2)_z(\boldsymbol{\sigma}_1 - \boldsymbol{\sigma}_2) + h^{(2)}(r)(\tau_1 \times \tau_2)_z(\boldsymbol{\sigma}_1 \times \boldsymbol{\sigma}_2)]. \quad (7.17)$$

Obviously, since the photon couples to the pion, this is a direct probe of pion exchange. Comparison of theory with experimental measurements

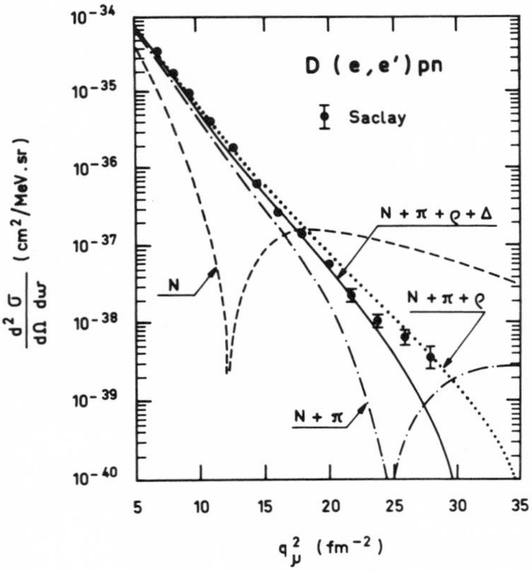

FIGURE 7.2. Shown is experimental data on the deuteron electro-disintegration cross section compared to predictions based on impulse approximation (dotted line) and impulse approximation supplemented by various meson exchanges (solid, dotted, and dash-dot lines). Exchange effects are clearly necessary for a good fit. (B. Frois, Nucl. Phys. *A434*, 57 (1985)).

is shown in Figure 7.2 and clearly reveals that inclusion of exchange currents—mesonic degrees of freedom within the nucleus—is absolutely necessary in order to obtain a good fit.

(2) *Elastic electron scattering.* A second test for meson exchange currents comes from recent elastic electron scattering measurements on light nuclear systems. Shown in Figure 7.3 is a comparison between experimental measurements of the magnetic form factor of ^{3}He as measured in electron scattering and theoretical predictions based upon impulse approximation only and impulse approximation plus meson exchange calculation [7]. It is clear that the diffraction minimum is moved from $q^2 = 8\ \text{fm}^{-2}$ to $q^2 = 18\ \text{fm}^2$ by mesonic effects and that this shift is *required* by the data. Thus it is fair to say that with respect to the electromagnetic interaction, "smoking gun" evidence for mesonic effects in the nucleus does exist. However, what about the weak interaction, which we have been emphasizing in this text? Do equally compelling arguments obtain in this case? Obviously, precise measurements will be required in order to answer this question, and the best such data exist in the arena of nuclear beta decay. Thus, I shall modify my question somewhat to ask, "Is there convincing evidence from nuclear beta decay for the existence of pionic degrees of

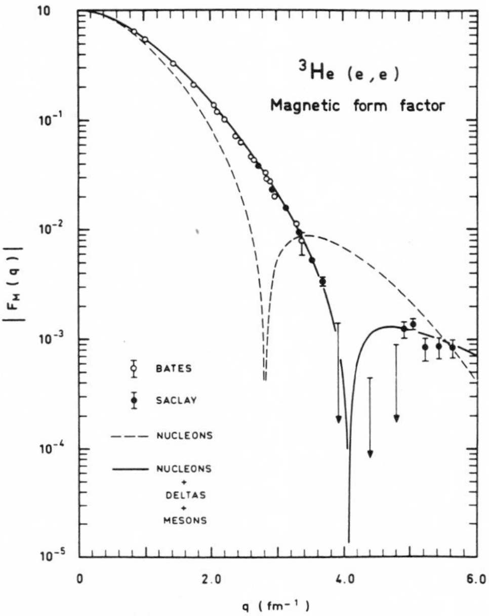

FIGURE 7.3. Shown is experimental data on the ^{3}He elastic magnetic (M1) form factor compared to predictions based on impulse approximation (dotted line) and impulse approximation plus meson exchange (solid line). Obviously, exchange effects are *required*. (B. Frois, Nucl. Phys. *A434*, ST (1985)).

freedom?" As we have previously seen, the interaction responsible for nuclear beta decay is

$$H_w = \frac{G}{\sqrt{2}} \cos \theta_c \langle \beta | V_\mu + A_\mu | \alpha \rangle l^\mu, \tag{7.18}$$

where, for allowed decay—$\Delta J = 0, \pm 1$, no parity change—one finds, in impulse approximation [8] (Figure 7.4),

$$\begin{aligned}
V_0^{(1)} &\sim g_V \sum_i \tau_i^{\pm} \\
\mathbf{V}^{(1)} &\sim g_M \sum_i \tau_i^{\pm} \boldsymbol{\sigma}_i \times \mathbf{q}/2m_N \\
A_0^{(1)} &\sim g_A \sum_i \tau_i^{\pm} i\boldsymbol{\sigma}_i \times \mathbf{L}_i \cdot \mathbf{q}/2m_N \\
\mathbf{A}^{(1)} &\sim g_A \sum_i \tau_i^{\pm} \boldsymbol{\sigma}_i,
\end{aligned} \tag{7.19}$$

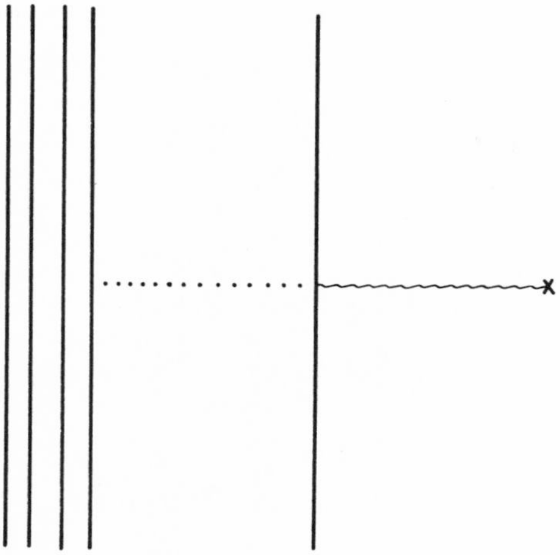

FIGURE 7.4. Interaction of the weak current J_μ with the nucleus in impulse approximation.

while the meson exchange contribution coming from Figure 7.5 is given, for the axial current, by the expressions [9]

$$\begin{aligned}
A_0^{(2)} &= f^2 g_A i(\tau_i \times \tau_j)^{\pm} \mathbf{q} \cdot (x_i \boldsymbol{\sigma}_j \cdot \hat{r} + x_j \boldsymbol{\sigma}_i \cdot \hat{r}) \\
&\quad \times \left(1 + \frac{1}{m_\pi r}\right) \frac{e^{-m_\pi r}}{4\pi r} \\
\mathbf{A}^{(2)} &= f^2 \frac{m_\pi}{2m_N} g_A(\tau_i \times \tau_j)^{\pm} \\
&\quad \times \left(\boldsymbol{\sigma}_j \cdot \hat{r}\boldsymbol{\sigma}_i \times \hat{r} - \boldsymbol{\sigma}_i \cdot \hat{r}\boldsymbol{\sigma}_j \times \hat{r} - \frac{2}{3}\boldsymbol{\sigma}_i \cdot \boldsymbol{\sigma}_j\right) \\
&\quad \times \left(1 + \frac{3}{m_\pi r} + \frac{3}{(m_\pi r)^2}\right) \frac{e^{-m_\pi r}}{4\pi m_\pi r},
\end{aligned} \tag{7.20}$$

with $f^2/4\pi = 0.08$ being the pseudovector $\pi - N$ coupling constant. It is then straightforward to construct the order of magnitude estimates:

$$\begin{aligned}
\frac{\mathbf{A}^{(2)}}{\mathbf{A}^{(1)}} &\sim f^2 \frac{m_\pi}{m_N} \left\langle \frac{e^{-m_\pi r}}{m_\pi r} \right\rangle \sim \text{few } \% \\
\frac{A_0^{(2)}}{A_0^{(1)}} &\sim f^2 \frac{m_N}{m_\pi} \left\langle \frac{e^{-m_\pi r}}{m_\pi r} \right\rangle \sim 50\%.
\end{aligned} \tag{7.21}$$

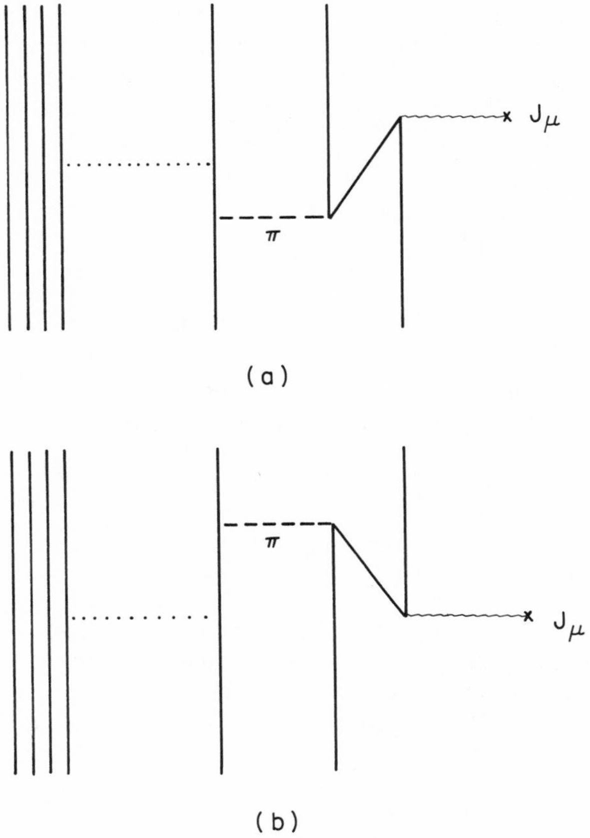

FIGURE 7.5. Interaction of the weak current J_μ with the nucleus via meson exchange.

Similarly we can show that

$$\frac{V^{(2)}}{V^{(1)}} \sim f^2 \frac{m_N}{m_\pi} \left\langle \frac{e^{-m_\pi r}}{m_\pi r} \right\rangle \frac{g_V}{g_M} \sim 10\%$$
$$\frac{V_0^{(2)}}{V_0^{(1)}} \sim f^2 \frac{m_\pi}{m_N} \left\langle \frac{e^{-m_\pi r}}{m_\pi r} \right\rangle \sim \text{few } \%. \tag{7.22}$$

Thus corrections to magnetic interactions should be significant, as already observed in the analogous example of electrodisintegration of the deuteron. However, exchange current contributions should be largest in matrix elements of A_0—the time component of the axial current. Unfortunately, from this point of view, allowed weak decays are usually dominated by the spatial component of the axial current—the Gamow-Teller. Terms involving A_0 are difficult—but not impossible—to detect, as I shall show.

The "secret" to this task is the observation that the nonrelativistic form of the nuclear matrix element of the weak axial vector current is given by [cf. eq. (4.8)]

$$l^\mu \langle \beta | A_\mu | \alpha \rangle = C^{M'k;M}_{J'1;J} \left(c \left(l_k + q_k \frac{1}{2M} l_0 \right) - d \left(q_0 l_k - q_k l_0 \right) \frac{1}{2M} + \dots \right). \tag{7.23}$$

Using the one-body results in eq. (7.19) yields the impulse approximation predictions for the Gamow-Teller and induced tensor form factors,

$$c = g_A \sum_i \tau_i^{\pm} \boldsymbol{\sigma}_i + \dots$$
$$d = g_A A \sum_i \tau_i^{\pm} i \boldsymbol{\sigma}_i \times \mathbf{L}_i \frac{1}{2m_N} + \dots . \tag{7.24}$$

We observe that the Gamow-Teller, induced tensor terms are sensitive primarily to spatial, timelike components of the axial current, respectively. Thus measurement of the induced tensor should provide a probe for the existence of meson exchange effects. We have already noted how the induced tensor can be detected. In section 4.2 I described correlation experiments involving mirror beta transitions that sought to demonstrate the existence or nonexistence of so-called second-class currents. This was accomplished by seeking a piece of the axial tensor coupling d, which changed sign between e^+ and e^- branches. Such a second class coupling was ruled out experimentally at the level of 10–20% of weak magnetism. At the time, it was noted that these correlation measurements also measured the component of the axial tensor whose origin was "induced" from ordinary (first-class) currents. We can now use these measurements to say something about mesonic degrees of freedom.

First, we list experimental results, as described more fully in section 4.2 [10],

	d_I/Ac
$A = 6$	0.2 ± 1.0
$A = 8$	1.9 ± 1.3
$A = 12$	3.8 ± 0.6
$A = 20$	5.2 ± 0.7,

(7.25)

and compare them with the values as calculated using only the impulse approximation (one-body) prediction:

$$\frac{d}{Ac} \simeq \frac{\langle f\|\sum_i \tau_i^{\pm} i\boldsymbol{\sigma}_i \times \mathbf{L}_i\|i\rangle + i\langle f\|\sum_i \tau_i^{\pm}(\boldsymbol{\sigma}_i \cdot \mathbf{r}_i\mathbf{p}_i + \mathbf{p}_i\boldsymbol{\sigma}_i \cdot \mathbf{r}_i)\|i\rangle}{\langle f\|\sum_i \tau_i^{\pm}\boldsymbol{\sigma}_i\|i\rangle} \tag{7.26}$$

Employing the p-shell wave functions of Cohen and Kurath and for $A = 20$ the S/D shell wave functions of Chung and Wildenthal, we find [11]

$$\begin{array}{ll} & (d/A_c)^{\mathrm{IA}} \\ A = 6 & 0.01 \\ A = 8 & 5.6 \\ A = 12 & 3.6 \\ A = 20 & 3.9. \end{array} \tag{7.27}$$

Comparison with experiment reveals that the theoretical predictions are consistent in size and sign, but only the cases $A = 6$, 12 agree with the values found experimentally. The question is whether inclusion of meson exchange effects improves matters.

A little thought indicates that there may exist some problems on this score, at least for the case of $A = 12$. Recall that the hand-waving estimate given in eq. (7.20) suggested that mesonic effects should provide an enhancement over the simple impulse approximation predictions by about 50% or so [12]. We can verify this by explicit calculation. Inclusion of pion exchange via the prescription

$$\left(\frac{d}{Ac}\right) = \left(\frac{d}{Ac}\right)^{\mathrm{IA}} + \left(\frac{d}{Ac}\right)^{\mathrm{MEC}} \tag{7.28}$$

with

$$\left(\frac{d}{Ac}\right)^{\mathrm{MEC}} = f^2 \frac{\langle f\|i(\boldsymbol{\tau}_i \times \boldsymbol{\tau}_j)^{\pm}(\mathbf{x}_i\boldsymbol{\sigma}_j \cdot \hat{r} + \mathbf{x}_j\boldsymbol{\sigma}_i \cdot \hat{r})\left(1 + \frac{1}{m_\pi r}\right)\frac{e^{-m_\pi r}}{m_\pi r}\|i\rangle}{\langle f\|\tau_i^{\pm}\boldsymbol{\sigma}_i\|i\rangle} \tag{7.29}$$

is found to yield

$$A = 12\colon \frac{d^{\mathrm{MEC}}}{Ac} \cong 1.3 \tag{7.30}$$

or

$$d^{\text{MEC}}/d^{\text{IA}} \sim 0.4, \tag{7.31}$$

as expected. Yet inclusion of both one- and two-body terms yields a prediction for the size of the tensor which lies far outside the limits permitted experimentally.

This point was troublesome for some time until it was recognized by Guichon and Samour [13] and by Morita [14] that core polarization effects can play an important role. By this I mean that the "true" wave function of a nucleon, which is nominally in a $1p$ level, also includes contributions from additional configurations, which represent higher levels of nuclear excitation. Since the simple shell model is reasonably successful, one hopes that this configuration mixing is small and may be treated perturbatively:

$$\begin{aligned} |\psi_{1p}\rangle &= |\chi_{1p}^{(+)}\rangle + \sum_n \frac{|\psi_n\rangle\langle\psi_n|V_r|\chi_{1p}^{(+)}\rangle}{E_{1p} - E_n} \\ &\equiv |1p\rangle + \varepsilon|\psi^{(r)}\rangle, \end{aligned} \tag{7.32}$$

where V_r is a residual interaction. Then we find, for example,

$$\begin{aligned} \langle\psi_{1p}|\tau\boldsymbol{\sigma}|\psi_{1p}\rangle &= \langle 1p|\tau\boldsymbol{\sigma}|1p\rangle + \mathcal{O}(\varepsilon^2) \\ \langle\psi_{1p}|i\tau\boldsymbol{\sigma}\times\mathbf{L}|\psi_{1p}\rangle &= \langle 1p|i\tau\boldsymbol{\sigma}\times\mathbf{L}|1p\rangle + \mathcal{O}(\varepsilon^2), \end{aligned} \tag{7.33}$$

so that such matrix elements are probably reliably calculated using the basic shell model prescription. However, for an operator such as $\boldsymbol{\sigma}\cdot\mathbf{r}\boldsymbol{\nabla}$, which also contributes to d_{I} [cf. eq. (7.26)], such corrections can be much more substantial:

$$\begin{aligned} &\langle\psi_{1p}|\sum_i i\tau_i(\boldsymbol{\sigma}_i\cdot\mathbf{r}_i\mathbf{p}_i + \mathbf{p}_i\boldsymbol{\sigma}_i\cdot\mathbf{r}_i)|\psi_{1p}\rangle \\ &\quad = \langle 1p|\sum_i i\tau_i(\boldsymbol{\sigma}_i\cdot\mathbf{r}_i\mathbf{p}_i + \mathbf{p}_i\boldsymbol{\sigma}_i\cdot\mathbf{r}_i)|1p\rangle + \mathcal{O}(\varepsilon) \end{aligned} \tag{7.34}$$

and can thereby affect the one-body predictions appreciably. Koshigiri et al. calculate using first-order perturbation theory [15],

$$d^{\text{core polarization}} = -1.2, \tag{7.35}$$

yielding a net result

$$\begin{aligned} d^{\text{Total}} &= d^{\text{IA}} + d^{\text{MEC}} + d^{\text{core polarization}} \\ &= 3.6 + 1.3 - 1.2 = 3.7, \end{aligned} \tag{7.36}$$

in good agreement with experiment. Similarly, Guichon and Samour find [13]

$$d^{\text{Total}} = 3.0 + 1.0 - 0.4 = 3.6. \tag{7.37}$$

The difference in numerical values here is due to use of different wave functions and residual interactions, but the conclusions are the same—only by inclusion of both meson exchange *and* core polarization effects can one obtain a consistent agreement with experiment.

We have recently extended these calculations to the remaining $A = 6$, 8, 20 systems and find similar results [16]:

$$\begin{aligned} A &= 6 \qquad d/Ac = 0.01 + 0.14 - 0.14 = 0.01 \\ A &= 8 \qquad d/Ac = 5.6 + 0.9 - 1.2 = 5.3 \\ A &= 20 \qquad d/Ac = 3.9 + 1.1 - 0.6 = 4.4. \end{aligned} \tag{7.38}$$

Comparison with experimental results given in eq. (7.24) reveals that while theory and experiment are in basic concurrence for $A = 6$ and $A = 20$, the size of the induced tensor predicted in the $A = 8$ case seems to be significantly larger than that seen experimentally. Recently announced $\beta\gamma$ correlation measurements in ^{20}F yielding [17]

$$d/Ac = 11.3 \pm 3.0 \tag{7.39}$$

could also prove troubling here.

I conclude that, although for the best-studied $A = 12$ system there does seem to exist evidence for the existence of a meson exchange effect at the expected—$\sim 50\%$—level, the situation in the remaining sectors is confusing at best. The combination of difficult experiments and model-dependent theoretical calculations—dependent on the form of residual interactions and the size of model spaces as well as the usual wave-function uncertainties—make definitive conclusions hard to come by. Nevertheless, the mesonic enhancement of the timelike component of the axial current is probably present but is shielded from direct observation by the opposite sign core polarization effects.

Thus, attempting to verify the existence of exchange currents via allowed nuclear beta decay is difficult at present. However, an alternative procedure exists, involving use of first forbidden beta decay between levels of opposite parity but identical spin. In this case the rank zero component of the axial current can participate and timelike and spatial pieces of A_μ are found to have comparable matrix elements. We find in impulse approximation

$$A_0^{[\text{Rank } 0]} \sim g_A \sum_i \tau_i^{\pm} \boldsymbol{\sigma}_i \cdot \mathbf{p}_i \frac{1}{2m_N}$$
$$\mathbf{A}^{[\text{Rank } 0]} \sim g_A \sum_i \tau_i^{\pm} \boldsymbol{\sigma}_i \cdot \mathbf{r}_i \mathbf{q}, \tag{7.40}$$

so that

$$\frac{\langle f|A_0|i\rangle}{\langle f|\mathbf{A}|i\rangle} \sim \frac{p_F}{q(m_N R)}. \tag{7.41}$$

Taking $p_F \sim 200$ MeV and $q \sim 10$ MeV as typical values, we find

$$\frac{\langle f|A_0|i\rangle}{\langle f|\mathbf{A}|i\rangle} \sim \mathcal{O}(1), \tag{7.42}$$

as claimed.

One can probe this rank zero axial current in various ways. The most straightforward is to utilize $0^+ - 0^-$ transitions, to which *only* the axial current contributes. The system that is best studied at present is that in $A = 16$ (cf. Figure 7.6), wherein both the beta decay,

$$^{16}\text{N}^*(0^-; 120 \text{ KeV}) \rightarrow {}^{16}\text{O} + e^- + \bar{\nu}_e,$$

and the inverse muon capture reaction,

$$\mu^- + {}^{16}\text{O} \rightarrow {}^{16}\text{N}^*(0^-; 120 \text{ KeV}) + \nu_\mu,$$

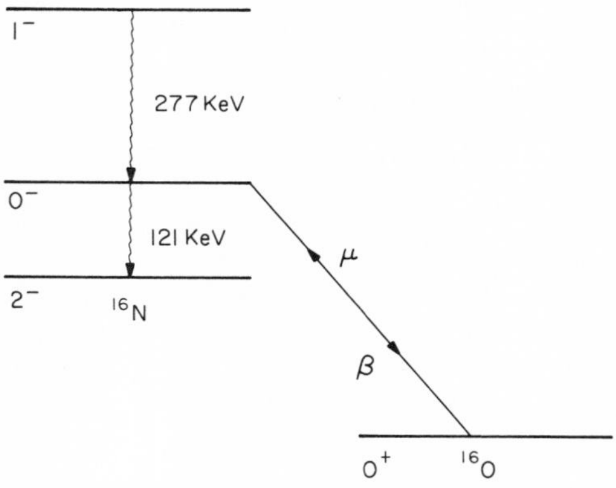

FIGURE 7.6. Level scheme for the $A = 16$ system.

have been studied experimentally. In the latter case there exists an accepted value of the capture rate [18],

$$\Gamma_\mu = 1560 \pm 94 \text{ sec}^{-1}. \tag{7.43}$$

However, in the case of the $0^- - 0^+$ beta decay reaction there exists at present some degree of uncertainty with the decay rate measured by the Osaka group:

$$\Gamma_\beta = 0.66 \pm 0.06 \text{ sec}^{-1} \qquad \text{Osaka [19]} \tag{7.44}$$

being significantly greater than that found elsewhere

$$\begin{aligned} \Gamma_\beta &= 0.497 \pm 0.023 \text{ sec}^{-1} && \text{Argonne [20]} \\ \Gamma_\beta &= 0.51 \pm 0.08 \text{ sec}^{-1} && \text{Louvain [21]} \\ \Gamma_\beta &= 0.480 \pm 0.024 \text{ sec}^{-1} && \text{Montreal [22].} \end{aligned} \tag{7.45}$$

Taking the ratio of muon capture to beta decay rates yields a value

$$\frac{\Gamma_\mu}{\Gamma_\beta} = \begin{cases} 3.2 \pm 0.2 \times 10^3 & \text{Argonne, Louvain, Montreal} \\ 2.4 \pm 0.3 \times 10^3 & \text{Osaka,} \end{cases} \tag{7.46}$$

which is somewhat insulated from overall normalization uncertainties.

Theoretical expressions for these rates are given by [23]

$$\begin{aligned} \Lambda_\mu &= 16.75 \times 10^3 |g_A{}^2(q^2)|^2 |M_\mu|^2 \\ \Lambda_\beta &= 21.3 |g_A{}^2(0)|\, |M_\beta|^2, \end{aligned} \tag{7.47}$$

where g_A is the axial coupling of neutron beta decay and

$$\begin{aligned} M_\mu &= \langle f| \sum_i \tau_i^-\left(j_0(k_\nu r_i) \frac{\boldsymbol{\sigma}_i \cdot \boldsymbol{\nabla}_i}{m_N} - g_\mu \frac{3}{k_\nu r_i} j_1(k_\nu r_i) \boldsymbol{\sigma}_i \cdot \mathbf{r}_i\right)|i\rangle \\ M_\beta &= \langle f| \sum_i \tau_i^+ \left(\frac{\boldsymbol{\sigma}_i \cdot \boldsymbol{\nabla}_i}{m_N} - g_\beta \boldsymbol{\sigma}_i \cdot \mathbf{r}_i\right)|i\rangle, \end{aligned} \tag{7.48}$$

with

$$\begin{aligned} g_\mu &= \frac{1}{3} k_\nu \left(1 - \frac{k_\nu}{2m_N}\left(\frac{g_P}{g_A} - 1\right)\right) \\ g_\beta &= -\frac{1}{3} E_0 \left(1 + \frac{3\alpha Z}{2E_0 R}\right). \end{aligned} \tag{7.49}$$

Theoretical evaluation of these matrix elements by Towner and Khanna yielded a result [24]

$$\left(\frac{\Lambda_\mu}{\Lambda_\beta}\right)^{\text{theo}} = 8 \times 10^3, \tag{7.50}$$

which is somewhat larger than found experimentally. Inclusion of meson exchange effects as done by these authors *does* indeed yield a reduction in the predicted muon capture to beta decay ratio. The precise value is sensitive to details about the residual interaction, nuclear wave functions, etc. However, values as low as

$$\left(\frac{\Lambda_\mu}{\Lambda_\beta}\right)^{\text{IA+MEC}} \sim 5 \times 10^3 \tag{7.51}$$

were obtained. These authors also noted that it is possible actually to fit the experimental number, provided one is willing to increase the value of the induced pseudoscalar coupling from the PCAC prediction,

$$(g_P/g_A)^{\text{PCAC}} = 7, \tag{7.52}$$

to a larger value,

$$(g_P/g_A) = 13 \pm 1. \tag{7.53}$$

However, in view of the theoretical uncertainties involved, I do not take this too seriously.

A second approach to this problem is that taken by Millener and Warburton [25], who deal with the beta-decay rates of several p-shell first forbidden decays. Besides the $0^- - 0^+$ ^{16}N transition discussed above, they also examine the following:

(1) the transition $^{11}\text{Be}(\frac{1}{2}^+;\text{g.s.}) \to {}^{11}\text{B}(\frac{1}{2}^-;2125\text{ KeV}) + e^- + \bar{\nu}_e$, wherein the rank-zero component can be separated from its rank-one counterpart by measurement of the electron-neutrino correlation. (The recoil ion energy was measured via detection of the Doppler shift associated with the subsequent deexcitation of the $^{11}\text{B}(\frac{1}{2}^+;2125\text{ KeV})$ state [26]);

(2) the transition $^{15}\text{C}(\frac{1}{2}^+;\text{g.s.}) \to {}^{15}\text{N}(\frac{1}{2};\text{g.s.}) + e^- + \bar{\nu}_e$, wherein the rank-zero component of the transition amplitude can be extracted by measurement of the spectral shape factor [27].

Parametrizing the decay rates as

$$\Lambda_\beta = \text{Phase Space}(\varepsilon_{\text{MEC}} M_\beta{}^T + a(Z,W_0) M_\beta{}^S)^2, \tag{7.54}$$

where $M_\beta{}^T, M_\beta{}^S$ are the impulse approximation matrix elements for timelike, spacelike axial currents, respectively, and ε_{MEC} is a phenomenological

TABLE 7.2
RANK ZERO AXIAL MATRIX ELEMENTS FOR VARIOUS P-SHELL BETA TRANSITIONS

Decay	M_β^T	$a(Z,E_0)M_\beta^S$	M_β^{exp}	ε_{MEC}
$^{11}Be(\frac{1}{2}^+) \to {}^{11}B(\frac{1}{2}^-)$	13.7	−9.7	13.0 ± 1.0	1.65
$^{15}C(\frac{1}{2}^+) \to {}^{15}N(\frac{1}{2}^-)$	28.1	−16.1	30.2 ± 0.4	1.65
$^{16}N(0^-) \to {}^{16}O(0^+)$	55.7	−29.0	58.3 ± 3.0	1.57

Shown are the results of calculations by Millener and Warburton [28] of the rank zero axial matrix elements for various p-shell beta transitions. Comparison with the experimental values given in column 4 yields the empirical meson exchange enhancement factors in the last column.

mesonic exchange enhancement factor associated with $M_\beta{}^T$, these authors find the results given in Table 7.2. Note that in each case a substantial—$\sim 60\%$—enhancement of the timelike axial current matrix element is required in order to fit experiment, just as suggested by eq. (7.21).

Before we conclude that these results on forbidden beta decay represent the smoking-gun evidence we have been seeking, it is important to note that considerable model dependence remains. For example, Towner and Khanna emphasized that for the ^{16}N transition, inclusion in a harmonic oscillator basis of two-particle, two-hole excitations to the ^{16}O ground state has the effect

$$M_\beta{}^S \to M_\beta{}^S(1+y) \quad \text{and} \quad M_\beta{}^T \to M_\beta{}^T(1-y),$$

with $y \sim 0.1$, thus suppressing the decay rate substantially (cf. Table 7.2). Similarly, Millener and Warburton have emphasized the importance of using realistic radial wave functions in evaluation of these first forbidden matrix elements. This is because such matrix elements,

$$\langle f|\boldsymbol{\sigma}\cdot\mathbf{r}|i\rangle, \qquad \langle f|\boldsymbol{\sigma}\cdot\boldsymbol{\nabla}|i\rangle,$$

are much more sensitive to details of the nuclear surface than are those such as the Gamow-Teller,

$$\langle f|\boldsymbol{\sigma}|i\rangle,$$

which dominate analysis of allowed decay. Warburton [27] has pointed out that in lowest order these corrections are overlapping, but in higher orders both types of effects must be carefully accounted for.

Finally, it is noteworthy that all the results described above were derived in the usual nonrelativistic formulation of nuclear physics. Recently both phenomenological considerations [29] and theoretical calculations [30] have suggested that use of a relativistic approach to nuclear structure may be useful. Thus, for example, in a simple mean-field approach to nuclear binding, the usual -50 MeV attractive nuclear potential consists of a strong $V_S \sim -400$ MeV attraction—due to exchange of the σ scalar boson—and a substantial $V_V \sim +350$ MeV repulsion—due to the exchange of the vector meson ω, which can be formulated in terms of a Dirac equation,

$$(\gamma_0(E - V_V) - \boldsymbol{\gamma} \cdot \mathbf{p} - m_N - V_S)\psi = 0. \tag{7.55}$$

Writing this in terms of an effective energy

$$E^{\text{eff}} = E - V_V \tag{7.56}$$

and effective mass

$$m_N^{\text{eff}} = m_N + V_S, \tag{7.57}$$

we see that an effective free particle Dirac equation can be written:

$$(\gamma_0 E^{\text{eff}} - \boldsymbol{\gamma} \cdot \mathbf{p} - m_N^{\text{eff}})\psi = 0, \tag{7.58}$$

with energy, mass parameters quite different from those typically employed. This shift should be particularly important for quantities such as the time-like axial current A_0, which involves lower components of the Dirac wave function. Defining

$$\psi = \begin{pmatrix} \psi_u \\ \psi_l \end{pmatrix}, \tag{7.59}$$

we find, using eq. (7.57),

$$\psi_l = \frac{\boldsymbol{\sigma} \cdot \mathbf{p}}{E^{\text{eff}} + m_N^{\text{eff}}} \psi_u. \tag{7.60}$$

Since

$$E^{\text{eff}} + m_N^{\text{eff}} = E + m_N - 750 \text{ MeV} \approx (E + m_N) \times 0.6, \tag{7.61}$$

such relativistic effects can be substantial. At the present time this subject is still controversial and we can draw no firm conclusions [31]. However, further study is clearly necessary in order to decide whether such relativistic considerations represent important corrections to the evaluation of matrix elements of the timelike axial current.

I conclude that, while a great deal of effort has gone into the attempt to answer the question I formulated at the beginning of this section—"Is there convincing evidence for the existence of mesonic exchange effects in nuclear beta decay?"—the jury is still out. Present evidence is clouded by various model dependencies. The "smoking gun" has not yet been found, even though a very strong circumstantial case *has* been constructed. Frankly, I personally am convinced, though not beyond the shadow of a doubt. The long and tortuous path we have followed in this section was necessary because we are not dealing with a symmetry principle, wherein many aspects of nuclear structure can be neglected, as shown in previous sections. Rather, we must face the reality that careful and detailed nuclear structure calculations are required, and that the current state of the art is almost but not quite able to provide a nearly model-independent answer.

References

[1] E. Feenberg, Rev. Mod. Phys. *19*, 239 (1947).
[2] J. D. Jackson, *Classical Electrodynamics*, Wiley, New York (1962), Ch. 4.
[3] R. J. Blin-Stoyle in *Mesons in Nuclei*, ed. M. Rho and D. H. Wilkinson, North-Holland, Amsterdam (1979), vol. I, p. 3.
[4] T.E.O. Ericson, Comm. Nuc. Part. Phys. *13*, 157 (1984).
[5] T.E.O. Ericson and M. Rosa-Clot, Ann. Rev. Nucl. Part. Sci. *35*, 271 (1985).
[6] D. O. Riska and G. E. Brown, Phys. Lett. *B38*, 193 (1972); J. F. Mothiot, Nucl. Phys. *A412*, 201 (1984).
[7] J. M. Cavedon et al., Phys. Rev. Lett. *49*, 986 (1982); P. Dunn et al., Phys. Rev. *C27*, 71 (1983).
[8] B. R. Holstein, Rev. Mod. Phys. *46*, 789 (1974).
[9] K. Kubodera et al., Phys. Rev. Lett. *40*, 755 (1978); R.A.M. Guichon et al., Phys. Lett. *B74*, 15 (1978).
[10] These values are extracted from data quoted in Ch. 4: $A = 8$, ref. 30; $A = 12$, ref. 31, and $A = 20$, ref. 32. For $A = 6$, I use the analysis given by W. E. Kleppinger, F. P. Calaprice, and B. R. Holstein, Nucl. Phys. *A293*, 46 (1977).
[11] J. F. Dubach, priv. comm.
[12] J. Delorme, Nucl. Phys. *374*, 541 (1982).
[13] P.A.M. Guichon and C. Samour, Nucl. Phys. *A382*, 461 (1982).
[14] M. Morita et al., Prog. Theor. Phys. Suppl. *60*, 1 (1976); M. Morita et al., Phys. Lett. *B73*, 17 (1978).

[15] K. Koshigiri et al., Prog. Theor. Phys. *66*, 358 (1981).
[16] J. F. Dubach and B. R. Holstein, unpublished.
[17] R. D. Rosa et al., Phys. Rev. *C37*, 2722 (1988).
[18] P. Guichon et al., Phys. Rev. *C19*, 987 (1979); F. R. Kane et al., Phys. Lett. *B45*, 292 (1973).
[19] T. Minamisono et al., Phys. Lett. *B130*, 1 (1983).
[20] H. R. Heath and G. T. Garvey, Phys. Rev. *C31*, 2190 (1985); C. A. Gagliardi et al., Phys. Rev. *C28*, 2423 (1983); C. A. Gagliardi et al., Phys. Rev. Lett. *48*, 914 (1982).
[21] L. Palffy et al., Phys. Rev. Lett. *34*, 212 (1975).
[22] L. A. Hamel et al., Z. Phys. *A321*, 439 (1985).
[23] I. S. Towner, Ann. Rev. Nucl. Part. Sci. *B6*, 115 (1986).
[24] I. S. Towner and F. C. Khanna, Nucl. Phys. *A372*, 331 (1981).
[25] D. M. Millener and E. K. Warburton, in *Nuclear Shell Models*, ed. M. Vallures and B. H. Wildenthal, World Scientific, Singapore (1985), p. 365.
[26] E. K. Warburton, D. E. Alburger, and D. H. Wilkinson, Phys. Rev. *C26*, 1186 (1982).
[27] D. E. Alburger, A. Gallman, and D. H. Wilkinson, Phys. Rev. *116*, 939 (1958).
[28] E. K. Warburton, in *Interactions and Structures in Nuclei*, ed. R. J. Blin-Stoyle and W. D. Wilkinson, IOP Publ. Inc., A. Hilgers (1988).
[29] J. M. Cameron, Nucl. Phys. *A434*, 261 (1985), Proc. LAMPF Workshop in Dirac Approaches to Nuclear Physics, LAMPF, Los Alamos (1985).
[30] J. D. Walecka, Ann. Phys. *83*, 491 (1974); B. D. Serot and J. D. Walecka, Adv. Nucl. Phys. *16*, 1 (1986); G. E. Brown, W. Weise, G. Baym, and J. Speth, Comm. Nucl. Part. Phys. *17*, 39 (1987).
[31] J. A. McNeil and J. R. Shepard, Phys. Rev. *C31*, 686 (1985), and *C33*, 1106 (1986); H.P.C. Rood, Phys. Rev. *C33*, 1104 (1986); G. Do Dang et al., Phys. Lett. *B153*, 17 (1985).

Chapter 8

SUMMARY

I have attempted in this monograph to indicate some of the ways in which weak interactions in nuclei are being utilized as a probe for not just interesting but also fundamental physics. Much of this work is right at the very frontiers of knowledge and its outcome may challenge our basic understanding of the electroweak interaction and of nature itself. Although I have discussed many topics, I have not really gone into trenchant detail on any subject and have presented only a rough outline in some cases. I have also omitted a great deal of material that could also fall into this category—for example, neutron-antineutron oscillations in matter, searches for the axion involving nuclear beta decay, tests for isotopic spin invariance of the nucleon-nucleon interaction, etc. These omissions are due to space limitations and the book's purpose. My goal has not been to present the subject in penultimate detail but rather to portray the field as a sketch of the landscape as of 1988. Peaks and valleys are included, but not the detail that would show up under finer resolution. The field is also changing. Between the time that I first presented this material in 1985 and wrote the final draft in 1988, many of the sections had to be rewritten, as new results in neutrino mass measurements, nuclear parity violation, and nuclear dipole moments had already changed our view of the world. And, of course, there was the supernova. This is as it should be. Physics is dynamic and evolving, and the picture I present is a bit blurred in many areas, indicating this motion. I hope that this overview will serve as a useful introduction to the subject and as a stimulus for others to continue to put things into focus.

INDEX